The Springer Series in Applied Machine Learning

Series Editor

Oge Marques, Florida Atlantic University, Boca Raton, FL, USA

Editorial Board

Over the last decade, Machine Learning and Artificial Intelligence methods and technologies have played an increasingly important role at the core of practical solutions for a wide range of tasks ranging from handheld apps to industrial process control, from autonomous vehicle driving to environmental policies, and from life sciences to computer game playing.

The Springer Series in "Applied Machine Learning" will focus on monographs, textbooks, edited volumes and reference books that provide suitable content and educate the reader on how the theoretical approaches, the algorithms, and the techniques of machine learning can be applied to address real world problems in a principled way.

The series starts from the realization that virtually all successful machine learning solutions to practical problems require much more than applying a set of standard off-the-shelf techniques. Such solutions may include principled ways of controlling parameters of algorithms, carefully developed combinations of standard methods, regularization techniques applied in a non-standard manner, probabilistic inference over the results of multiple models, application-driven statistical analysis of results, modeling of the requirements for robustness, or a mix of practical 'tricks' and thoroughly analyzed approaches. In fact, most practical successes rely on machine learning applied at different levels of abstraction of the tasks that are addressed and the books of this series will provide the reader a basis for understanding how these methods were applied, why the techniques were chosen, what are the pitfalls, and how the reader can build a solution for her/his own task.

Our goal is to build a series of state of the art books for the libraries of both machine learning scientists interested in applying the latest techniques to novel problems, and subject matter experts, technologists, or researchers interested in leveraging the latest advances in machine learning for developing solutions that work for practical problems. Our aim is to cover all topics of relevance for applied machine learning and artificial intelligence.

The series will also publish in-depth analyses and studies of applications to real world problems in areas such as economics, social good, environmental sciences, transportation science, manufacturing and production planning, logistics and distribution, financial planning, structural optimization, water resource planning, network design, and computer games.

In addition, this series will aim to address the interests of a wide spectrum of practitioners, students and researchers alike who are interested in employing machine learning and AI methods in their respective domains. The scope will span the breadth of machine learning and AI as it pertains to all application areas both through books that address techniques specific to one application domain and books that show the applicability of different types of machine learning methods to a wide array of domains.

Poornachandra Sarang

Thinking Data Science

A Data Science Practitioner's Guide

 Springer

Poornachandra Sarang
Practicing Data Scientist & Researcher
Mumbai, India

ISSN 2520-1298   ISSN 2520-1301   (electronic)
The Springer Series in Applied Machine Learning
ISBN 978-3-031-02365-1   ISBN 978-3-031-02363-7   (eBook)
https://doi.org/10.1007/978-3-031-02363-7

This Springer imprint is published by the registered company Springer Nature Switzerland AG
The registered company address is: Gewerbestrasse 11, 6330 Cham, Switzerland

Preface

Chapter 1 (Data Science Process) introduces you to the data science process that is followed by a modern data scientist in developing those highly acclaimed AI applications. It describes both the traditional and modern approach followed by a current day data scientist in model building. In today's world, a data scientist has to deal with not just the numeric data, but he needs to handle even text and image datasets. The high-frequency datasets are another major challenge for a data scientist. After this brief on model building, the chapter introduces you to the full data science process. As we have a very large number of machine learning algorithms, which can apply to your datasets, the model development process becomes time consuming and resource intensive. The chapter introduces you to AutoML that eases this model development process and hyper-parameter tuning for the selected algorithm. Finally, it introduces you to the modern approach of using deep neural networks (DNNs) and transfer learning.

Machine learning is based on data, more the data that you have; it makes learning better. Let us consider a simple example of identifying a person in a photo, video, or just in real life. If you have a better knowledge or have more features of that person known to you, the identification becomes a simple task. However, in machine learning, the machine does not like this. In fact, we consider having many features a curse of dimensionality. This is mainly due to two reasons—we, human-beings, cannot visualize data beyond three dimensions and having many dimensions demands enormous resources and training times. Chapter 2 (Dimensionality Reduction) teaches you several techniques for bringing down the dimensions of your dataset to a manageable level. The chapter gives you an exhaustive coverage of dimensionality reduction techniques followed by a data scientist.

After we prepare the dataset for machine learning, the data scientist's major task is to select an appropriate algorithm for the problem that he is trying to solve. The Classical Algorithms Overview (Part I) gives you an overview of the various algorithms that you will study in the next few chapters.

The model development task could be of a regression or classification type. Regression is a well-studied statistical technique and successfully implemented in machine learning algorithms. Chapter 3 (Regression Analysis) discusses several regression algorithms, starting with simple linear to Ridge, Lasso, ElasticNet, Bayesian, Logistic, and so on. You will learn their practical implementations and how to evaluate which best fits for a dataset.

Chapter 4 (Decision Trees) deals with decision trees—a fundamental block for many machine learning algorithms. I give in-depth coverage to building and traversing the trees. The projects in this chapter prove their importance for both regression and classification problems.

Chapter 5 (Ensemble: Bagging and Boosting) talks about the statistical ensemble methods used to improve the performance of decision trees. It covers both bagging and boosting techniques. I cover several algorithms in each category, giving you definite guidelines on when·to use them. You will learn several algorithms in this chapter, such as Random Forest, ExtraTrees, BaggingRegressor, and BaggingClassifier. Under boosting, you will learn AdaBoost, Gradient Boosting, XGBoost, CatBoost, and LIghtGBM. The chapter also presents a comparative study on their performances, which will help you in taking your decisions on which one to use for your own datasets.

Now, we move on to classification problems. The next three chapters cover K-Nearest Neighbors, Naive Bayes, and Support Vector Machines used for classification.

Chapter 6 (K-Nearest Neighbors) describes K-Nearest Neighbors, also called KNN, which is the simplest and starting algorithm for classifications. I describe the algorithm in-depth along with the effect of K on the classification. I discuss the techniques of obtaining the optimal K value for better classifications and finally provided guidelines on when to use this simple algorithm.

Chapter 7 (Naive Bayes) describes Naive Bayes' theorem and its advantages and disadvantages. I also discuss the various types, such as Multinomial, Bernoulli, Gaussian, Complement, and Categorical Naive Bayes. The Naive Bayes is useful in classifying huge datasets. A trivial project toward the end of the chapter brings out its importance.

Now, we come to another important and widely researched classification algorithm, and that is SVM (Support Vector Machines). Chapter 8 (Support Vector Machines) gives an in-depth coverage to this algorithm. There are several types of hyperplanes that divide the dataset into different classes. I fully discuss the effects of the kernel and its various types, such as Linear, Polynomial, Radial Basis, and Sigmoid. I provide you with definite guidelines for kernel selection for your dataset. SVM also requires tuning of its several parameters, such as C, Degree, Gamma, and so on. You will learn parameter tuning. Toward the end, I present a project that shows how to use SVM and concludes with SVM's advantages and disadvantages in practical situations.

A data scientist need not have a deep knowledge of how these algorithms are designed? Having only a conceptual understanding of the purpose for which they were designed suffices. So, in Chaps. 3 to 8, I focus on explaining the algorithm's concepts, avoiding mathematics on which we base them and giving more importance to how we use them practically.

Now comes the next challenge for a data scientist, and that is clustering a dataset without having labeled data points. We call this unsupervised learning. I have a huge section (Part II) comprising Chaps. 9 through 16 for clustering, giving you an in-depth coverage for several clustering techniques. The notion of cluster is not well-defined and usually there is no consensus on the results produced by clustering algorithms. So, we have lots of clustering algorithms that deal with small, medium, large, and really huge spatial datasets. I cover many clustering algorithms explaining their applications for various sized datasets.

Chapter 9 (Centroid-Based Clustering) discusses the centroid-based clustering algorithms, which are probably the simplest and are the starting points for clustering huge spatial datasets. The chapter covers both K-Means and K-Medoids clustering algorithms. For K-Means, I describe its working followed by the explanation of the algorithm itself. I discuss the purpose of objective function and techniques for selecting optimal clusters. These are called Elbow, Average Silhouette, and the Gap Statistic. This is followed by a discussion on K-Means limitations and where to use it? For the K-Medoids algorithm, I follow a similar approach describing its working, algorithm, merits, demerits, and implementation.

Chapter 10 (Connectivity-Based Clustering) describes two connectivity-based clustering algorithms: Agglomerative and Divisive. For Agglomerative clustering, I describe the Simple, Complete, and Average linkages while explaining its full working. I then discuss its advantages, disadvantages, and practical situations where this algorithm finds its use. For Divisive clustering, I take a similar approach and discuss its implementation challenges.

Chapter 11 (Gaussian Mixture Model) describes another type of clustering algorithm where the data distribution is of Gaussian type. I explain how to select the optimal number of clusters with a practical example.

Chapter 12 (Density-Based Clustering) focuses on density-based clustering techniques. Here I describe three algorithms—DBSCAN, OPTICS, and Mean Shift. I discuss why we use DBSCAN? After discussing preliminaries, I discuss its full working. I then discuss its advantages, disadvantages, and implementation with the help of a project. To understand OPTICS, I first explain to you a few terms like core distance and reachability distance. Like the earlier one, I discuss its implementation with the help of a project. Finally, I describe Mean Shift clustering by explaining its full working and how to select the bandwidth. A discussion on the algorithm's strengths, weaknesses, applications, and a practical implementation illustrated with the help of a project follows this.

Chapter 13 (BIRCH) discusses another important clustering algorithm called BIRCH. This is an algorithm that helps data scientists in clustering huge datasets, where all the earlier algorithms fail. BIRCH splits the huge dataset into subclusters by creating a hierarchical tree-like structure. The algorithm does clustering incrementally eliminating the need for loading the entire dataset into memory. In this chapter, I discuss why and where to use this algorithm and explain its working by showing you how the algorithm constructs a CF tree.

Chapter 14 (CLARANS) discusses another important algorithm for clustering enormous sized datasets. This is called CLARANS. The CLARA (Clustering for

Large Applications) is considered an extension to K-Medoids. CLARANS (Clustering for Large Applications with Randomized Search) is a step further to handle spatial datasets. This algorithm maintains a balance between the computational cost and the random sampling of data. I discuss both CLARA and CLARANS algorithms and present you a practical project to understand how to use CLARANS.

Chapter 15 (Affinity Propagation Clustering) describes an altogether different type of clustering technique which is based on gossiping. This is called Affinity Propagation clustering. I fully describe its working by explaining to you how we use gossiping for forming groups having affinity toward each other and their leader. This algorithm does not require you to have a prior estimation of the number of clusters. I explain the concept of responsibility and availability matrices while presenting its full working. Toward the end, I demonstrate its implementation with a practical project.

Toward the end of the clustering section, you will learn two more clustering algorithms: these are STING and CLIQUE. Chapter 16 (STING & CLIQUE) discusses these algorithms. We consider them both density and grid based. The advantage of STING lies in the fact that it does re-scan the entire dataset while answering a new query. Thus, unlike previous algorithms, this algorithm is computationally far less expensive. The STING stands for STatistical INformation Grid. I explain how the algorithm constructs the grid and uses it for querying. CLIQUE is a subspace clustering algorithm that uses a bottom-up approach while clustering. This algorithm provides better clustering in the case of high-dimensional datasets. Like earlier chapters, I present the full working of the algorithms, their advantages, disadvantages, and their practical implementations in projects.

After discussing the several classical algorithms, which are based on statistical techniques, we now move on to an evolutionally technique of machine learning and that is ANN (Artificial Neural Networks). The ANN technology definitely brought a new revolution in machine learning. Part III (ANN: Overview) provides you an overview of this technology.

In Chap. 17 (Artificial Neural Networks), I introduce you to ANN/DNN technology. You will first learn many new terms like back-propagation, optimization/loss functions, evaluation metrics, and so on. I discuss how to design an ANN architecture on your own, how to train/evaluate it, and finally how to use a trained-model for inferring unseen data. I introduce you to various network architectures, such as CNN, RNN, LSTM, Transformers, BERT, and so on. You will understand the latest Transfer Learning technology and learn to extend the functionality of a pre-trained model for your own purposes.

Chapter 18 (ANN-Based Applications) deals with two practical examples of using ANN/DNN. One example deals with text data and the other one with image data. Besides other things like designing and training networks, in this chapter, you will learn how to prepare text and image datasets for machine learning.

With the wide choices of algorithms and selection between classical and ANN approach of model building, we make the data scientist's life quite tough. Fortunately, the researchers and engineers have developed tools to help data scientists in the above selections.

Chapter 19 (Automated Tools) talks about the automated tools for developing machine learning applications. The modern tools automate almost all workflows of model development. These tools build efficient data pipelines, select between GOFAI (Good Old Fashioned AI—classical algorithms) and ANN technologies, select the best performing algorithm, ensemble models, design an efficient neural network, and so on. You just need to ingest the data into such tools and they come up with the best performing model on your dataset. Not only this, some also spill out the complete model development project code—a great boon to data scientists. This chapter gives a thorough coverage of this technology.

The last chapter, Chap. 20 (Data Scientist's Ultimate Workflow), is the most important one. It merges all your lessons. In this chapter, I provide you with a definite path and guidelines on how to develop those highly acclaimed AI applications and become a Modern Data Scientist.

The entire book at the end will make you a most sought-after data scientist. For those of you who are currently working as a data scientist, this book will help you become a modern data scientist. A modern data scientist can handle numeric, text, and image datasets, is well conversant with GOFAI and ANN/DNN development, and can use automated tools including MLaaS (Machine Learning as a Service).

So, move on to Chap. 1 to start your journey toward becoming a highly skilled modern data scientist.

Mumbai, India   Poornachandra Sarang

Contents

Part III   ANN: Overview

Chapter 1
Data Science Process

With lucrative and millions of job offerings in data science, there are many who aspire to be a data scientist. The assumption is you need to have a college degree in statistics, computer science, or engineering. However, today even the nontechnicals thrive for a data scientist job. With the proper learning path, it is possible for most of you to become a data scientist, irrespective of your college education. This chapter starts with the data science process and sets the path for your goal of becoming a data scientist. If you already hold a position of a data scientist, the book will provide you guidelines to become a "modern" data scientist.

The first step toward becoming a data scientist is understanding the data science process. What briefly is the role of a modern data scientist? How has he developed the AI applications with an outstanding success that we use in our everyday life? We use face tagging, Alexa, and autonomous cars—these are AI applications. As far as businesses are concerned, how do they use AI? Tech giants like Amazon build their own recommendation systems for product selling. Banks use AI for fraud detection and early warnings. Even governments use AI for surveillance, traffic monitoring, and so on. The applications are many. Modern data scientists developed all such applications.

I have been using the adjective "modern" for a reason. Today, with the technology change and the industry's new demands, the role of a data scientist has changed appreciably. Traditionally, data scientists use the well-studied statistical solutions to develop predictive machine learning models. They developed such models on mostly numeric data available in our traditional databases—relational, non-relational. Even if the data contained few character columns, we would convert them into numeric, as the machine understands only the binary format. Number of such columns and the number of data points in the early days of machine learning was low enough for the machines in those days to handle. Today, the industry requirements rely on developing machine learning (ML) models on an enormous corpus of text and image data. Developing such models requires enormous computing resources and an altogether different data preprocessing compared to numeric data taken from traditional databases. Also, there are many technological advances

P. Sarang, *Thinking Data Science*, The Springer Series in Applied Machine Learning, https://doi.org/10.1007/978-3-031-02363-7_1

in model building itself. Becoming a modern data scientist, you need to understand how to handle these new data types and learn the modern technologies of machine learning.

So, let us begin our journey by first understanding the data science process. I will first introduce you to the traditional model building process, followed by old-hat data scientists. No, do not take it wrong. Though these processes were developed many years ago, they still find their usefulness in modern data science. I will provide you with definite guidelines on when to use the traditional approach and when to use a more-advanced modern approach.

Traditional Model Building

All these years, a data scientist, while building an AI application, would first start with the exploratory data analysis (EDA). After all, understanding the data yourself is very vital in telling the machine what it means. In technical terms, it is important for us to understand the features (independent variables) in our dataset to do a predictive analysis on the target, a dependent variable. Using these features and targets, we would create a training dataset for training a statistical algorithm. Such EDA, many-a-times, requires a deep domain knowledge. And that is where people having domain knowledge in various vertical industries thrive to become a data scientist. I will try to meet the aspirations of every such individual.

As said earlier, in those days, the data was mostly numeric. As a data scientist, you have to make sure that the data is clean for feeding it to an algorithm. So, there comes data-cleansing. For this, one has to first find out if there are any missing values. If so, either remove those columns from your analysis or impute the proper values in those missing fields. After we ensure the data is clean, you need to do some preprocessing on it to make it ready for machine learning.

The various steps required in data preparation are studying the data variance in columns, scaling data, searching for correlations, dimensionality reduction, and so on. For this, he would use many available tools for data exploring, get visual representations of data distributions and correlations between columns, and use several dimensionality reduction techniques. The list is endless; process is time-consuming and laborious.

After the data scientist makes the dataset ready for machine learning, his next task is to select an appropriate algorithm based on his knowledge and experience. After the algorithm is trained, we say we have built the model. Data scientist now uses any known performance evaluations methods to test the trained model on his test datasets. If the performance metrics do not give acceptable accuracies, he will try tweaking the hyper-parameters of the algorithm. If that does not work, he may have to go back to the data preparation stage, selecting new features, do additional features engineering and further dimensionality reductions, and then retrain his algorithm for an improved accuracy. If this too does not work out, he will try another statistical algorithm. The entire process continues over many iterations until he creates a fine-tuned model on his dataset. You see, the entire process is not just time-consuming, but also requires a good amount of knowledge of statistical methods. You must understand several classical machine learning algorithms and have a deep knowledge of EDA techniques. That is to say, knowledge width rather than a deep knowledge of each algorithm is required in developing efficient machine learning models.

Now, we look at the modern approach.

Modern Approach for Model Building

The traditional process that I have described above is highly laborious and time-consuming. Not only this, as a data scientist you would need an excellent knowledge of statistics as a subject, several tools available for EDA, learn several machine learning algorithms, and know the various techniques for evaluating the model's performance. What if somebody automates this entire process? Such automation has been available for many years. As a data scientist, you need to gain these new skills.

We have auto-sklearn that works on the popular sklearn machine learning library that does both regression and classification automated model building. Not just the model selection, it has built-in functionality to build a better performing data pipeline created by applying Bayesian optimization. It also uses statistical ensemble techniques to do model aggregations. If you decide to use an artificial neural network (ANN) approach for model building, we have AutoKeras that outputs the best performing network design for your dataset. Several commercial and non-commercial tools are available in this space—just to mention a few, H2O.ai, TPOT, MLBox, PyCaret, DataRobot, DataBricks, and BlobCity AutoAI. The last one provides an interesting feature that, to my knowledge and at the time of this writing, nobody else does. It generates a complete .ipynb file (project source), which, as a data scientist, you can claim as your own invention. After all, all clients ask for the originality and the source code for the project.

I have included a full chapter on AutoML for you to master this novel approach.

The next important aspect of modern machine learning development is the introduction of new data types, and that is text and image data.

AI on Image Datasets

With the outstanding success in developing predictive models using statistical techniques, we also call it classical ML and sometimes even use the term "good old-fashioned AI" (GOFAI); the industry wanted to apply their learnings on image data initially for object detection. It probably started with detection of a certain class of objects in an image, and then the techniques were extended to classify dogs and cats, face tagging, and identifying persons and finally even to perform such operations in real-time data streams. Today, we have machine learning applications like traffic analysis that work on live data streams. All these new models made machine learning engineers (in the short term, we call them ML engineers) and data scientists to develop and learn new methods for processing image data. Finally, our computers understand only binary data; though image data is binary data, we still need to transform this data into a machine-understandable format. Note that each single image comprises several binary data items (pixels) and each pixel in an image is a RGB representation in an image data file.

The older classical ML technology failed to meet these new requirements, and the industry started looking at the alternative approaches. The ANNs (artificial neural networks) technology, which was invented many decades ago, came to the rescue. The modern computing resources made the use of this technology workable in developing such models. The neural network training requires several gigabytes of memory, GPUs, and many hours of training. Tech giants had those kinds of resources, and they trained networks, which we can reuse and extend their functionalities for our own purpose. We call this new technology Transfer Learning. I have given an exhaustive coverage later in the book on this new technology—transfer learning.

Model Development on Text Datasets

After seeing the success of ANN/DNN (deep neural networks) technology in building image applications, researchers started exploring its application to text data. There came the new term and the field of its own—natural language processing (NLP). Preparing text data for machine learning requires a different kind of approach as compared to numeric. Numeric data is contained in databases having a few columns. Each column is a potential candidate for a feature. Thus, in numeric machine learning models, the number of features is typically very low.

Now, consider the text data, where each word or a sentence is a potential candidate for a feature. For email-spam applications, you use every word in your text as a feature, while for a document summarization model, you will use every sentence as a feature. Considering the vocabulary of words that we have, you can easily imagine the number of features. The traditional dimensionality reduction techniques that we have for numeric data would not apply for a text dataset in reducing its dimensions. You need to cleanse the entire text corpus. For cleansing text data, you would need several steps, removing punctuations, stop words, removing/converting numbers, lowercasing, and so on. Further to this data cleaning operation, to reduce the features count, we need to apply a few more techniques like building a dictionary of unique words, stemming, lemmatization, and so on. Finally, you need to understand tokenization so that these words/sentences would be transformed into machine-understandable formats.

For text data, the context in which a word appears also plays an important role in its analytics. So, there came a new branch called natural language understanding

(NLU). So, we introduced many new concepts like bag-of-words (BoW), tf-idf (Term Frequency Inverse Document Frequency), bi-grams, n-grams, and so on. The researchers developed many new ANN architectures over the years for creating text-based AI applications, beginning with RNN (Recurrent Neural Networks), followed by LSTM (Long Short-Term Memory), and currently the transformers and its different implementations. Today, we have many models based on the transformer architecture. BERT (Bidirectional Encoder Representations from Transformers) is one of the most widely used one. Then, we have models like GPT for text generation. The list is endless.

Model Building on High-Frequency Datasets

High-frequency datasets is another challenging requirement that has come up in the industry these days. Why do I say challenging? Here, the dataset frequency keeps changing over time—the data is dynamic. So, your entire data science process, that includes building efficient data pipelines, features engineering, and model training/evaluation, has to be dynamic too. Businesses like Amazon and Jio (a large telecom company in India) collect several terabytes of data every day. Processing and doing data analytics on such high volume, high-frequency datasets is not a simple task.

Having seen the requirements and needs for developing modern AI applications, I will now show you the workflow that a modern data scientist follows to develop those highly acclaimed ML models.

Data Science Process

The first step in model building is to prepare the data for algorithm/ANN training.

Data Preparation

The data in this world comes in two formats—numeric or character based. A database table may contain both numeric and character fields. A character-based data is also available in a text document. When you do data analytics using machine learning, the data may be in a relational database, may be structured or unstructured, or may even come from a NoSQL database source. Also, a data scientist is required to work on a large text corpus, such as news or novels. If we pick the data from the Web, it will contain HTML tagging. We require a data scientist to work on image datasets too.

Both numeric and text-based data require a different processing treatment. Let us first consider numeric data.

Fig. 1.1 Processing
numeric fields

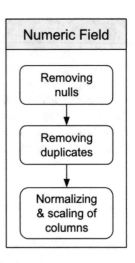

Numeric Data Processing

Using numeric data comes in application developments like weather forecasting, predicting sales during Christmas, deciding on the bid price for your house during an auction, and so on.

The workflow required in processing numeric data is given in Fig. 1.1.

As you know, several times, the database may contain null values in a numeric field. If you have a sizable amount of data available for machine learning, you may decide to delete rows containing null fields. However, if you do not have enough data points, you will usually replace such null fields with their mean/median values. If the database is not indexed, you may find duplicate rows in a table; you will remove such duplicates.

Each column in the database may have a varied distribution, and across columns, the min-max range may differ. Thus, all columnar data must be normalized and scaled to the same scale. Typically, the data is scaled to a range of −1 to +1 or 0 to 1, for better training of the algorithm.

Now, I will talk about what kind of processing is required for text-based data.

Text Processing

The preprocessing of the text data depends on the type of application that you are trying to develop. A character field in a database column may contain values like Male and Female to show the gender. If this field is important for your model, you must replace Male with 0 and Female with 1, say. That is to say, we must convert such categorical data into a numeric value as the computers understand only binary data and do not understand the characters. We call this conversion as encoding. There are two types of encoding we use: LabelEncoder or One-Hot Encoder.

Now, there is another kind of ML model development that requires a different treatment for character-based data. Consider an email spam detection application. If the email contains certain words like "lottery," "winner," etc., it may be treated spam. Thus, an email-client application must classify emails containing such words into the spam category, while we treat others as non-spam. For this, someone must tokenize the entire email text into words to detect the occurrences of such words. We would consider each word in the text as a feature of machine learning. Just imagine the number of words and thus the number of features a document would provide. Typically, for the dataset available in a database, we do not come across this kind of situation as each column in a database is considered as a feature for machine learning and the number of columns is usually a low number and is easily manageable.

Thus, processing of text data for applications like spam detection, text summarization, and movie-rating requires lots of processing steps as compared to numeric data. We call such processing as natural language processing (NLP), or in some advanced applications like language translation, we need to apply natural language understanding (NLU) techniques. These days, you also see applications such as *fake news* generation that require a good size of vocabulary—many features to deal with.

Preprocessing Text Data

A general workflow for processing text data is depicted in Fig. 1.2.

First, we remove all punctuation marks from the text. Why so? Because even punctuation marks would be tokenized and added to our list of features. Then, the text contains several stop words like "the," "this," "at," and so on. Such words make little sense when you do document categorization or sentiment analysis. So, we remove such stop words from the text.

After this basic processing, we move a step forward toward our next goal of feature extraction. For this, we tokenize the document to extract words or sentences. You would tokenize the text into sentences when you want to summarize a passage. I will now briefly describe word tokenization. When you tokenize the text into words, you may find the occurrence of words like "run," "running," and "ran"; such words convey only one meaning that is running, which is represented as a single keyword "run" that is added to our features repository. We call this process stemming. I just gave you an example here. To process raw text, you need several more techniques like Lemmatization, POS tagging, Vectorization, Chunking, Chinking, and so on. The NLP is an altogether different ball game, and I will describe this preprocessing separately in Chap. 18 (ANN-Based Applications).

Our next workflow is to understand the data that you have just cleaned up.

Fig. 1.2 Processing text
corpus

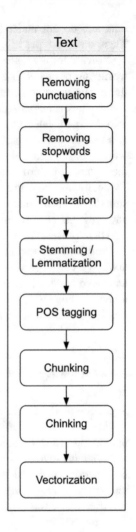

Exploratory Data Analysis

Figure 1.3 shows the workflow for exploratory data analysis (EDA) for understanding the data.

To understand the data, there are several libraries available for data visualization. The most commonly used Python libraries are Matplotlib and Seaborn that can provide a very advanced level of visualization. You may do scatter plots, histograms,

Fig. 1.3 Data visualization
libraries

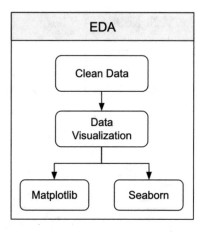

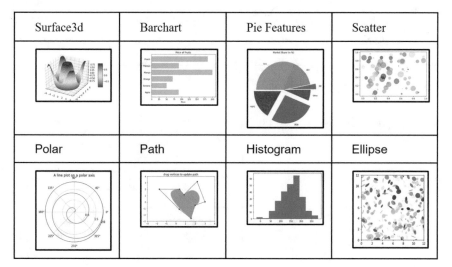

Fig. 1.4 Matplotlib sample outputs

bar charts, pie charts, three-dimensional plots, contours, polar plots, and whatever
else you need to understand the data. Figure 1.4 shows some capabilities of
Matplotlib.

Figure 1.5 shows some advanced plots created by Seaborn.

Such data visualizations help in detecting outliers, understanding correlations,
data distributions, and so on. We use this information for further refining our training
datasets.

Next, we come to feature engineering.

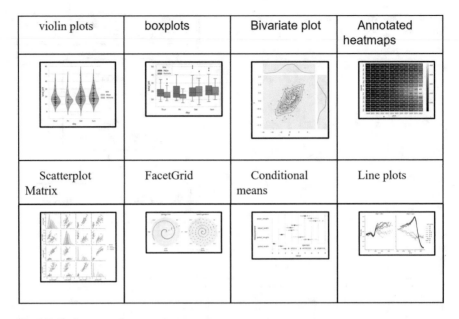

violin plots	boxplots	Bivariate plot	Annotated heatmaps
Scatterplot Matrix	FacetGrid	Conditional means	Line plots

Fig. 1.5 Seaborn sample outputs

Fig. 1.6 Options in features engineering

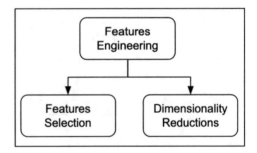

Features Engineering

Selecting appropriate features while keeping the features count low is very important in training the model efficiently. We call this features engineering. The features engineering workflow has two paths, features selection and dimensionality reduction. This is shown in Fig. 1.6.

Feature selection means selecting and excluding certain features without changing them. Dimensionality reduction reduces the features count by reducing the features dimensions using many techniques developed by researchers over years.

With domain knowledge, you can eliminate the unwanted features by manual inspection and simply dropping those columns from the database. You may remove features that have missing values and low variance or are highly correlated. You may use techniques such as univariate feature selection or recursive feature elimination. You may even use sklearn's *SelectFromModel* to remove features with values below a set threshold.

To reduce features, you may also use some dimensionality reduction techniques. One such widely used technique is PCA, that is, Principal Component Analysis. This is especially useful in cases with excessive multicollinearity, or explanation of predictors is not on your priority list.

Features engineering is a complex subject. Do not worry about it now. I will explain this in greater detail in a later chapter—called dimensionality reduction.

After you are done with the features and target selection, you are ready to build the model.

Deciding on Model Type

We classify a machine learning problem into the three major categories as depicted in Fig. 1.7. I am not including clustering here, which needs a special treatment. I have devoted a full section comprising several chapters on clustering.

When you want to predict a certain value, target as you call it in ML terms, you develop a regression model. When you want to group a dataset into a known or unknown number of groups, you use classification. A customer segmentation model falls into this category. If you are developing an AI model to play chess, or rather to play any game that requires ongoing learning, you develop a reinforcement learning model. As I do not cover the game development in this book and other applications that use reinforcement learning, I will not further discuss this reinforcement learning model. Now, let us focus on the regression and classification models, where most of the industry requirements lie. Here comes the next workflow.

Fig. 1.7 Selecting model based on the task

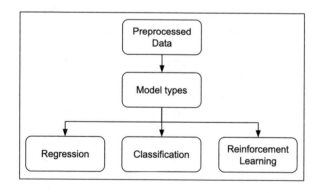

Fig. 1.8 Supervised/
unsupervised learning

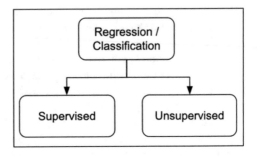

Model Training

For both regression and classification models, the model training can be a supervised learning or unsupervised learning, as seen in Fig. 1.8.

In case of supervised learning, you must have a labeled dataset, so to say that for each data point in the dataset, the target value is known. Using this labeled dataset, the model will tune its hyper-parameters and get ready to infer an unknown data point. You may test the model's accuracy on a validation dataset, which is part of your original labeled dataset, but not used during training.

In case of unsupervised learning, you do not have labeled data, or rather, it is impossible to create a labeled dataset because of its size. In such cases, you will use those machine learning algorithms which will do the data analysis of your dataset on their own. To cite an example here, an object detection model like OpenCV and YOLO was trained using unsupervised learning.

Algorithm Selection

The major daunting task for a data scientist is to decide which algorithm to use. For both regression and classification problems, there are many algorithms available in our repositories. The challenge is to select the one which is most suitable for the dataset and which can achieve a very high accuracy while predicting an unseen data point. This book will help you understand several algorithms and select an appropriate one for your application. To give you a quick overview of the several algorithms available for machine learning, look at Fig. 1.9.

It is important for a data scientist to understand these algorithms, if not full implementation, at least the concepts behind it. If you understand the algorithm conceptually and its purpose, then you would know which one to use for your current need. After all, there is somebody who has efficiently implemented these algorithms, and we may not do a better job than those hard-cored developers. Even if the implementation is not fully optimized, it is still okay in most cases. If your model is doing real-time inferences, then only this high-level optimization would be called

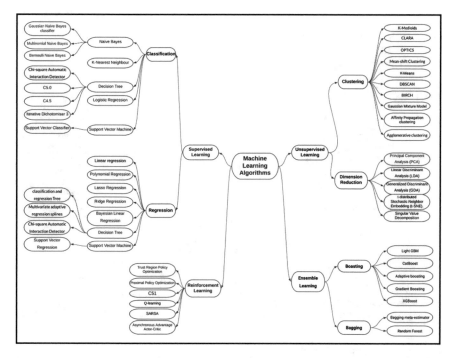

Fig. 1.9 Exhaustive list of classical ML models

for. Trust me, almost all the libraries which are out there in the market provide optimized coding.

Even for DNN, there is a vast repository of architectures and algorithms available, which are summarized in Fig. 1.10.

Now, the question that arises is if I use another ML algorithm to solve the same problem, will I get better accuracy? The answer to this question is non-trivial, and many get frustrated in trying out different algorithms manually on the same dataset until they find out the best performing one. Fortunately, the technology has advanced, and there is help available to you in selecting the algorithm. We call it AutoML.

AutoML

Figure 1.11 depicts the AutoML process.

Several AutoML libraries and frameworks, both free-to-use and commercial, are available in the market. One of the popular open-source libraries is auto-sklearn. These libraries/frameworks provide automatic data preparation, selecting the best performing machine learning model on the dataset and tuning hyper-parameters of the selected algorithm. In short, they cover the full machine learning process that I

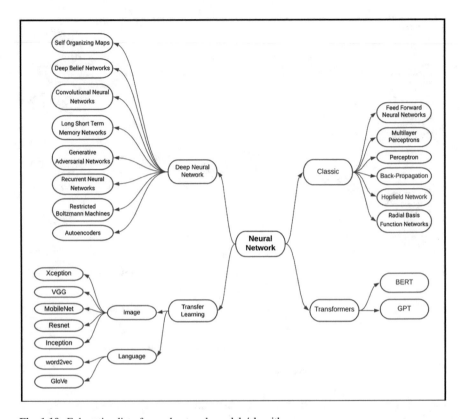

Fig. 1.10 Exhaustive list of neural network models/algorithms

have not even fully described so far. All you need to do is to provide them with a dataset, and they will suggest you the best performing model, all fine-tuned and ready-to-use for production. They provide the model rankings based on their accuracy scores, and you can decide which one to use for your purpose.

These days you would find offerings from Google, Microsoft, and Amazon who provide machine learning as a service (MLaaS), where you just need to upload your dataset on their servers and eventually download a model pipeline which can be hosted and used via a web service. Though this is fascinating, a true data scientist can out-beat these automated services. Second, most of these follow a black-box approach and do not provide you with the source code of model development for you to fiddle with. A data scientist usually uses an automated service to quickly narrow down his search on model selection.

For your own purposes and development, you may use some free libraries like auto-sklearn, AutoKeras, or commercial versions of H2O.ai. I have provided an exhaustive list of such frameworks in the AutoML chapter. These libraries support many ML models for training and evaluation. They will train each model on your dataset. After the training, they evaluate the model's performance. Finally, they rank

Fig. 1.11 AutoML process

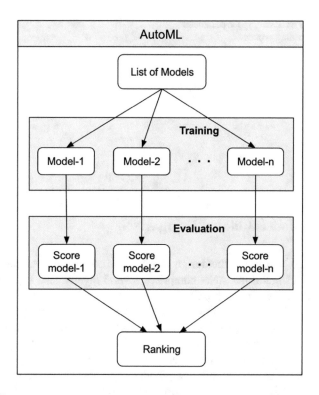

different models based on their accuracy scores. Isn't this nice? Maybe you struggle to believe this. Wait until I show you an actual implementation in a later chapter. Not only the algorithm selection, they also provide parameter fine-tuning and ultimately provide you with a machine learning model that is ready to use. Some AutoML libraries (BlobCity) also spill out the full source code for model building.

Hyper-Parameter Tuning

The next workflow as seen in Fig. 1.12 is to fine-tune the hyper-parameters of the selected algorithm.

After deciding on the algorithm to use, the next step is to train it on a training dataset. Each algorithm uses several parameters, which need to be fine-tuned. We call this hyper-parameter tuning, which is an iterative process. We can check the tuning by examining the model's accuracy score on the validation dataset. For hyper-parameter optimization, you may use frameworks like Optuna. After you achieve your desired level of accuracy, the model is ready for production use.

So far, the workflows I have shown help you in creating a machine learning model based on traditional statistical-based modeling. We also call this GOFAI (Good Old-Fashioned Artificial Intelligence). Next, I will define a workflow for developing models based on neural networks.

Fig. 1.12 Hyper-parameter
tuning process

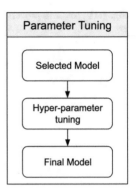

Model Building Using ANN

As you know, the design of ANN comes from the structure of a human brain. Like our brain, the ANN comprises several layers, each layer containing several nodes (neurons in our brain). We show a typical ANN architecture in Fig. 1.13.

Your job while designing an ANN architecture is to decide on the number of layers, the type of connections between the layers, and the number of nodes in each layer. We show the complete workflow for ANN development in Fig. 1.14.

After you define the architecture, feed the data to the network; let it propagate through the network and produce some results. Based on these results, the network will do the back propagation to fine-tune the hyper parameters (the weights assigned to each link). This is again an iterative process and after several training iterations, so-called epochs in ML terms, the network will provide you with a good level of accuracy. If the accuracy scores are not acceptable, try feeding in more data for training so that the network understands your data better. If this does not work, go back to the drawing board and adjust the number of layers, number of nodes in each layer, the optimization algorithms and the error functions that you have previously set for network learning, and so on. I will describe this entire process when I discuss the model development based on ANN in later chapters.

Though you may design your own ANN architecture for small applications and medium-sized datasets, creating a huge architecture is not simple. Such architectures are resource intensive and require training times spanning over several weeks. Image recognition and object detection models like R-CNN and YOLO were developed on complex architectures requiring enormous resources and training times. Such networks are called deep neural networks. Can we use their pre-trained models for our own purpose? The answer to this question is a big *yes*.

Models Based on Transfer Learning

To understand the term transfer learning, let me take an example. We have pre-trained image classification models that can detect a dog in an image. What if I want to know about the dog's breed? Such models, as they detect hundreds of

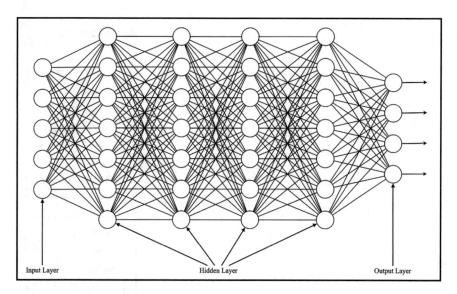

Fig. 1.13 ANN/DNN schematic

Fig. 1.14 ANN
development process

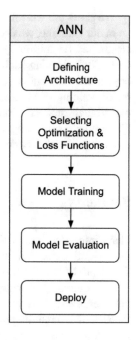

different objects, rarely go into such minor classifications. We can now extend their pre-trained model by adding our own network layers and train only these additional layers for determining the dog's breed. And this is what we call transfer learning.

Fig. 1.15 Transfer learning
process

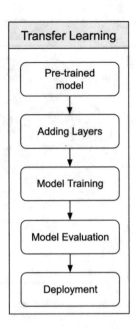

We show the workflow for transfer learning model development in Fig. 1.15.

In our example, we use the pre-trained model for object recognition. Such models are trained on millions of images and have taken up many weeks of training on GPU/TPUs. Having such processing power and time for training at your disposal is beyond the scope of many of us. Only the tech giants have infrastructures to do such model training. Fortunately, they allow us to use their pre-trained models. We can extend their models, add a few more layers, and train them for our further purpose, like predicting the dog's breed after the model detects a dog in an image. I will provide you with a concrete example of how this is done later in the book.

Summary

I have introduced you to the complete data science process, followed by an advanced data scientist. For building efficient ML models, you need to develop several skills. The depth of each skill is usually shallow. To be a successful data scientist, what it really requires is the knowledge width and not the depth. Finally, a data scientist is not a researcher to develop new algorithms or visualization techniques. You need to focus on understanding the concepts behind these technologies and how to use them in building your own models. What is important is the decision *which one to use where*, and that's what this book is going to guide you into. Keep reading!

Chapter 2
Dimensionality Reduction

Creating Manageable Training Datasets

After cleansing your dataset, performing EDA, and preparing your dataset for machine learning, your next task is to reduce its dimensions. In machine learning, the dataset size and dimensions both are equally important considerations while training the algorithm. Even if your dataset has manageable dimensions, you should still drop the dimensions, such as correlated columns, for better learning of the algorithms. The dimensionality reduction is a key to creating manageable training datasets, and that's what you will learn in this chapter.

In a Nutshell

A machine will learn better with more data; however, having too much data can also be problematic. Here are some issues that you would like to consider:

- A dataset having many features is difficult to explore.
- It is extremely difficult to visualize data in over three-dimensional space.
- Larger datasets imply more training times and resources.
- Correlated features cause an additional burden on training.

Considering all such factors, we prefer to reduce the dimensions of the dataset without losing the vital information in the data. We call this dimensionality reduction.

Let me first elaborate on why we need to reduce dimensions of a dataset.

Why Reduce Dimensionality?

One would say having more features for inference is better than having just a handful of features. In machine learning, this is not true. Having many features rather adds to the woes of a data scientist. We call it the "curse of dimensionality." A high-dimensional dataset is considered a curse to a data scientist. Why so? I will mention a few challenges a data scientist faces while handling high-dimensional datasets:

- A large dataset is likely to contain many nulls, so you must do thorough data cleansing. You will either need to drop those columns or impute those columns with appropriate data values.
- Certain columns may be highly correlated. You just need to pick up only one among them.
- There exists a high probability of over-fitting the model.
- Training would be computationally expensive.

Looking at these issues, we must consider ways to reduce the dimensions; rather, use only those really meaningful features in the model development. For example, having two columns, like *birthdate* and *age*, are meaningless as both convey the same information to the model. So, we can easily get rid of one without a compromise in the model's ability to infer unseen data. An *id* field in the dataset is totally redundant for machine learning and should be removed. Such manual inspections for reducing the dimensions would be practically impossible, and thus, many techniques are developed for dimensionality reduction programmatically.

Dimensionality Reduction Techniques

Dimensionality reduction essentially means reducing the number of features in your dataset. Two approaches can do this—keeping only the most important features by eliminating unwanted ones and the second one by combining some features to reduce the total count.

Lot of work has been done on dimensionality reduction, and we have devised several techniques for the same. You need to learn all these techniques as one technique may not meet our purpose; most of the time you will need to apply several techniques in a certain order to achieve the desired outcome. I will now discuss the following techniques. Though the list is not complete, it is surely exhaustive:

- Columns with missing values
- Filtering columns based on variance
- Filtering highly correlated columns
- Random forest
- Backward elimination
- Forward features selection
- Factor analysis

- Principal component analysis
- Independent component analysis
- Isometric mapping
- t-distributed stochastic neighbor embedding (t-SNE)
- Uniform Manifold Approximation and Projection (UMAP)
- Singular value decomposition (SVD)
- Linear discriminant analysis (LDA)

To illustrate these techniques, I have created a trivial project, which is discussed next.

Project Dataset

The project is based on the loan eligibility dataset. It has 12 features on the basis of which the customer's eligibility for a loan sanction is decided. Information on the dataset is shown in Fig. 2.1.

The dataset is designed for classification. The target is *Loan_Status*, which is either approved or disapproved. The loan status depends on the values of other 11 fields (*Loan_ID* is not useful to us), which are the features for our experimentation on dimensionality reduction. We try to reduce these dimensions and see its impact on our target variable. Thus, we start with our first DR technique.

```
df.info()
```

```
<class 'pandas.core.frame.DataFrame'>
RangeIndex: 614 entries, 0 to 613
Data columns (total 13 columns):
 #   Column             Non-Null Count  Dtype
---  ------             --------------  -----
 0   Loan_ID            614 non-null    object
 1   Gender             601 non-null    object
 2   Married            611 non-null    object
 3   Dependents         599 non-null    object
 4   Education          614 non-null    object
 5   Self_Employed      582 non-null    object
 6   ApplicantIncome    614 non-null    int64
 7   CoapplicantIncome  614 non-null    float64
 8   LoanAmount         592 non-null    float64
 9   Loan_Amount_Term   600 non-null    float64
 10  Credit_History     564 non-null    float64
 11  Property_Area      614 non-null    object
 12  Loan_Status        614 non-null    object
dtypes: float64(4), int64(1), object(8)
memory usage: 62.5+ KB
```

Fig. 2.1 Loan eligibility dataset structure

Columns with Missing Values

In a large dataset, you may find a few columns with lots of missing values. When you do a count of null value checks on each column, you will know which columns have missing values. You either impute the missing values or eliminate those columns with the assumption that they will not affect the model's performance.

To eliminate the specific columns, we decide on a certain threshold value; say, if the column has over 20% missing values, eliminate it. We can do this in Python, with a simple code like this:

```
df.isnull().sum()/len(df)
```

```
a = df.isnull().sum()/len(df)
variables = df.columns[:-1]
variable = []
for i in range(0,len(df.columns[:-1])):
    if a[i]>0.03:    #setting the threshold as 3%
        variable.append(variables[i])
print(variable)
```

This is the output on our dataset:

```
'Self_Employed', 'LoanAmount', 'Credit_History']
```

As you see, the columns *Self_Employed*, *LoanAmount*, and *Credit_history* are dropped as the number of missing values in these columns cross our set threshold limit.

In some situations, you may still decide to keep those columns having missing values. In this case, you will need to impute the missing values with either the mode or the median value of that column. In our dataset, let us consider the three columns, loan amount, loan amount term, and credit history that have missing values. We will plot the histograms of the data distribution for these three columns, which is shown in Fig. 2.2.

Looking at the histograms, you know that loan amount term and credit history have discrete distributions, so we will impute the column's mode value into the missing values. For the loan amount, we will use the median for the missing values. We show this in the code snippet below:

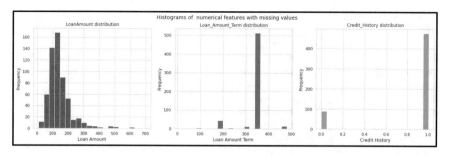

Fig. 2.2 Histograms of a few numerical features

```
df['LoanAmount'].fillna(
        df['LoanAmount'].median(),
        inplace=True)
df['Loan_Amount_Term'].fillna(
        df['Loan_Amount_Term'].mode()[0],
        inplace=True)
df['Credit_History'].fillna(
        df['Credit_History'].mode()[0],
        inplace=True)
```

Next, we will check for the variance in each column.

Filtering Columns Based on Variance

The columns that show low variance add little to the model training and can be safely dropped. To check the variance of the columns, we ensure all columns are numeric. For this, we will first convert our categorical columns into numeric ones. After this, you can check the variance in each column by calling the *var* method on the dataframe.

```
df.var()
```

This is the output on our dataset:

```
Gender                1.778751e-01
Married               2.351972e-01
Dependents            1.255589e+00
Education             1.708902e-01
Self_Employed         2.859435e-01
ApplicantIncome       3.732039e+07
CoapplicantIncome     8.562930e+06
LoanAmount            7.074027e+03
Loan_Amount_Term      4.151048e+03
Credit_History        1.241425e-01
Property_Area         6.201280e-01
Loan_Status           2.152707e-01
dtype: float64
```

As you observe, the variance for the *Credit_History* is very low, so we can safely drop this column. For datasets having large numbers of columns, this kind of physical examination of variance may not be practical, so you may set a certain threshold for filtering out those columns, as you did with detecting columns having large numbers of missing values. This is done using the following code snippet:

```
# Omitting 'Loan_ID' and 'Loan_Status'
numeric = df[df.columns[1:-1]]
var = numeric.var()
numeric_cols = numeric.columns
variable = []
for i in range(0, len(numeric_cols)):
  if var[i]>=10:    # variance threshold
    variable.append(numeric_cols[i])
variable
```

This is the output:

```
['ApplicantIncome', 'CoapplicantIncome', 'LoanAmount',
'Loan_Amount_Term']
```

As you see, the above-listed four columns exhibit lots of variance, so they are significant to us in machine training. You may likewise filter out the columns having low variance and thus insignificant to us.

Filtering Highly Correlated Columns

When two features approximately carry similar information, such as birth date and age, we say that they are highly correlated. Having correlated columns adds to the algorithm's complexity and sometimes leads to over-fitting.

You plot the correlation matrix to examine the correlations between the various features. Figure 2.3 shows such a matrix for our dataset.

This kind of plot is also called a heat map. If non-diagonal values are higher than a certain threshold, one of the corresponding variables can be safely removed.

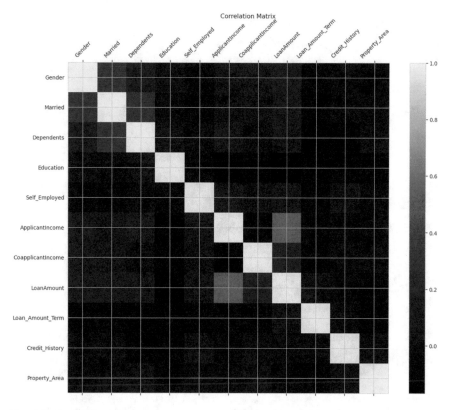

Fig. 2.3 Correlation matrix

You may also do the correlation check through a program code like this:

```
feature_cols = df.columns[1:-1]
corr_values = df[feature_cols].corr()


indexes = np.tril_indices_from(corr_values)

for coord in zip(*indexes):
    corr_values.iloc[coord[0], coord[1]] = np.NaN

corr_values = (corr_values
                .stack()
                .to_frame()
                .reset_index()
                .rename(columns={'level_0':'feature1',
                                 'level_1':'feature2',
                                 0:'correlation'}))


corr_values['abs_correlation'] =
                 corr_values.correlation.abs()


corr_values
```

The partial output on our dataset is shown in Fig. 2.4.

	feature1	feature2	correlation	abs_correlation
0	Gender	Married	0.336094	0.336094
1	Gender	Dependents	0.149674	0.149674
2	Gender	Education	0.024382	0.024382
3	Gender	Self_Employed	-0.025022	0.025022
4	Gender	ApplicantIncome	0.094472	0.094472
5	Gender	CoapplicantIncome	0.073308	0.073308

Fig. 2.4 Correlations in our dataset

You may also generate a plot of absolute correlation values as seen in Fig. 2.5.

You can then get a list of column pairs for which the absolute correlation is above a certain threshold. You will use this information to drop one of the correlated columns.

You may also set filters to isolate the column pairs having correlation index crossing a specified value.

```
corr_values.sort_values(
        'correlation', ascending=False).query(
        'abs_correlation>0.5')
```

The output is shown in Fig. 2.6.

As you see, the income and loan amount are relatively highly correlated, so you may use just one of the two. In huge datasets, the threshold can be around 0.90 or 0.95.

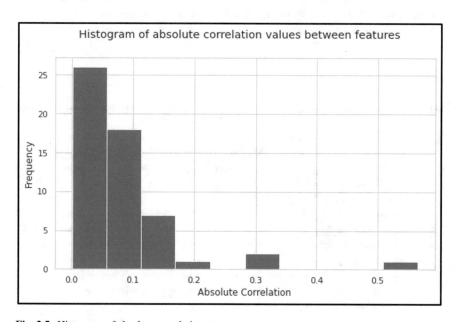

Fig. 2.5 Histogram of absolute correlations

	feature1	feature2	correlation	abs_correlation
41	ApplicantIncome	LoanAmount	0.565181	0.565181

Fig. 2.6 Columns having absolute correlation greater than 0.5

So far, you studied the feature selection techniques based on manual inspection of datasets. I will now show you a technique that somewhat eliminates this manual process.

Random Forest

Random forest is very useful in feature selections. The algorithm produces a feature importance chart, as seen in Fig. 2.7. This chart helps you in eliminating features having a low impact on model's performance.

Looking at the above chart, we can drop the low importance features such as *Education, Gender, Married,* and *Self_Employed.*

We widely use random forest in features engineering because of its in-built features importance package. The *RandomForestRegressor* computes a score based on a feature's impact on the target variable. The visual representation of these scores makes it easier for a data scientist to create a final list of features. As an example, looking at the above figure, a data scientist may select only the first four features as they have the maximum impact on the target for model building.

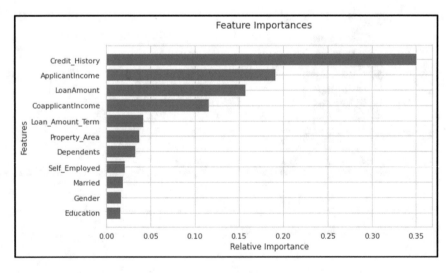

Fig. 2.7 Plot of features importance generated by random forest

Backward Elimination

In backward elimination, we train the model on all the n variables and then keep eliminating one feature at a time that has the least effect on the model's performance until they leave us with no more features to drop. The algorithmic steps are:

1. Train the model on the full dataset.
2. Remove one feature at random and retrain.
3. Repeat step 2 for all $n-1$ features.
4. Remove the feature from the set that has the least influence on the model's performance.
5. Repeat steps 2–4 until there are no more features left.

We use *sklearn.feature_selection.RFE* algorithm to implement backward elimination. The *RFE* eliminates the features of lesser significance recursively. It determines the importance of each feature through any specific attribute. One parameter to *RFE* is a supervised learning estimator, such as logistic regression. The estimator provides information on the feature importance. We repeat the process until you reach a desired number of features.

Let us consider we want to get the four most significant features in our dataset. We do this by using the following code snippet:

```
X = df.drop('Loan_Status', axis=1)
y = df['Loan_Status']
estimator = LogisticRegression(max_iter=150)
selector = RFE(estimator,
               n_features_to_select = 4,
               step = 1)
selector.fit(X, y)
X_selected = selector.transform(X)
```

Note that we used logistic regression as our estimator in the above code. After the *RFE* runs fully, you will get the list of supported features by checking the *support_* attribute value.

```
print('Support features')
print(selector.support_)
```

This is the output in my run:

```
Support features
[ True True False True False False False False False True False]
```

As per this output, the first, second, fourth, and tenth features, which are marked as True, are the most significant ones. You can also get the ranking of each feature by examining the *ranking_* attribute.

```
print('Features ranking by importance')
print(selector.ranking_)
```

This is the output in my run:

```
Features ranking by importance
[1 1 4 1 3 8 7 6 5 1 2]
```

A ranking value of 1 shows higher significance. Thus, features at index 0, 1, 3, and 9 are most significant in our case. You can print the names of these four most significant features by writing a small function:

```
def get_top_features():
 rank_1=[]
 for i in range(0,len(selector.ranking_)):
   if selector.ranking_[i]==1:
     rank_1.append(i)
 print('The four most informative features are:')
 print(X.iloc[:,rank_1].columns)
```

Calling this function gave the following output:

```
The four most informative features are:
Index(['Gender', 'Married', 'Education', 'Credit_History'],
dtype='object')
```

You observe that *Gender, Married, Education,* and *Credit_History* are the four most significant features for our model training.

Forward Features Selection

The forward features selection follows the exact opposite of the backward elimination process. We start with the single most important feature and then keep on adding more to it recursively until we do not observe any significant improvement in the model's performance. We use the *SelectKBest* function as shown here for selecting top seven features:

```
X = df.drop('Loan_Status', axis=1)
y = df['Loan_Status']
X_selected = SelectKBest(f_classif, k=7).fit(X, y)
```

The parameter *f_classif* to the *SelectKBest* function is a function that takes two arrays *X* and *Y* and returns either a pair of arrays (scores, p-values) or an array of scores. This function works only for classification tasks.

We transform the p-values into scores using the following statement:

```
scores = -np.log10(X_selected.pvalues_)
```

You may then plot the features importance as seen in Fig. 2.8.

As earlier, you can get a list of top features by writing a small function as shown here:

```
def get_best_features(scores):
  ind = np.argpartition(scores, -7)[-7:]
  print('The seven best features are:')
  print(X.iloc[:,ind].columns)
```

Calling this function, gave me the following output:

```
The seven best features are:
Index(['Loan_Amount_Term', 'Property_Area', 'LoanAmount',
'CoapplicantIncome',
  'Married', 'Credit_History', 'Education'],
  dtype='object')
```

Again, you may try a different *k* value in the *SelectKBest* function to get a more or a smaller number of significant features depending on your needs.

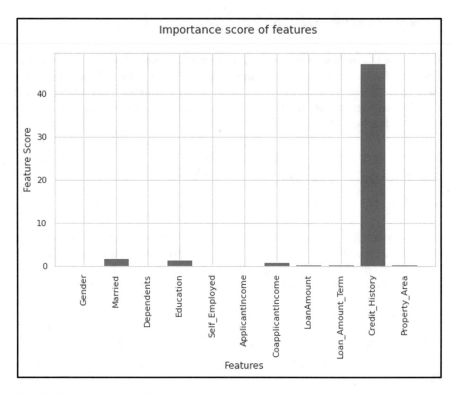

Fig. 2.8 Feature scores for all features

Factor Analysis

In factor analysis, we arrange all the features in groups such that within a group, the features are highly correlated, but they have a low correlation with features of other groups. For example, features like income and spend are highly correlated; higher-income people usually spend more. We put them into a group, which is also called a *factor*. We define an important term which is *eigenvalues*. This corresponds to the total amount of variance that a component can explain. A factor with eigenvalue greater than 1 explains more variance. So, we select only those features with values greater than 1.

To show the use of factor analysis, I used a Python module called *factor_analyzer* that is built for exploratory and factor analysis (EFA). It performs EFA using a minimal residual (MINRES), a maximum likelihood (ML), or a principal factor (PF) solution. You install the module using pip install as follows:

```
!pip install factor_analyzer
```

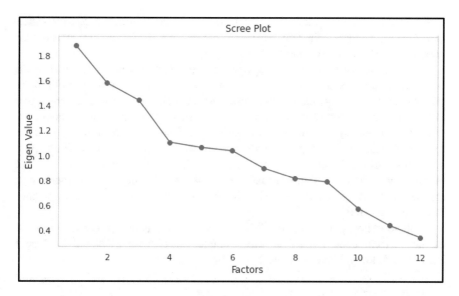

Fig. 2.9 Scree plot of factor analysis

You apply the analyzer on your dataset using the following code:

```
from factor_analyzer import FactorAnalyzer
fa = FactorAnalyzer(rotation = None,
                            impute = "drop",
                            n_factors=df.shape[1])
fa.fit(df)
```

After factoring, obtain the eigenvalues.

```
ev,_ = fa.get_eigenvalues()
```

A plot of eigenvalues versus factor, also called "Scree Plot" is seen in Fig. 2.9.
You observe that there are six components that have eigenvalue > 1. You will use these components for training your model.

Principal Component Analysis

Principal component analysis (PCA) is a well-known statistical process used in EDA and predictive modeling. It converts the correlated features into a set of linearly uncorrelated features using orthogonal transformations. The newly transformed features are called the *principal components*. The principal components are the

eigenvectors of the covariance matrix constructed on the dataset. It is closely related
to the factor analysis that you have seen earlier and is the simplest of the true
eigenvector-based multivariate analyses.

PCA has found a wide range of applications in many domains. A few real-world
applications are image processing, movie recommendation engines, optimizing
power allocation, and so on. It finds its usefulness in EDA and creating the predictive
models. During exploratory data analysis, it is difficult to visualize the data in over
three dimensions. We use PCA to reduce a multidimensional data to two or three
dimensions while preserving the data's variation as much as possible, so that we can
plot a multidimensional dataset into a 2-D or 3-D chart. For predictive model
building, we reduce a high-dimensional dataset, say having 100 dimensions to
maybe 30 dimensions without losing much of its variance, that to say, without
sacrificing the vital information.

I will now show you how to use PCA on our loan dataset that you have used in all
the above techniques. I will use sklearn implementation for PCA and the kernel
PCA. The following code snippet shows how it is applied:

```
X = df.drop('Loan_Status', axis=1)
y = df['Loan_Status']


pca = PCA(n_components=2)
X_pca = pca.fit_transform(X)


kpca = KernelPCA(n_components=2, kernel='rbf',
                 gamma=15, random_state=42)
X_kpca = kpca.fit_transform(X)
```

After creating the principal components, you can visualize them using a scatter
plot. This is shown in Fig. 2.10.

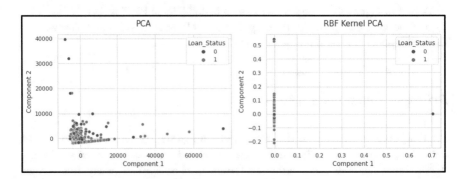

Fig. 2.10 PCA and kernel PCA

As you observe, Kernel-PCA has given us better separation. The normal PCA could not efficiently transform the nonlinear data into a linear form.

As PCA is a widely used technique for both EDA and features reduction, I have created a more elaborate project to illustrate its use.

PCA on Huge Multi-columnar Dataset

The project that I am going to describe now uses a dataset having 64 features and one target. Having 64 columns, EDA on this dataset would be difficult. We will apply PCA to see if we can get some visualization after reducing the dataset to two or three dimensions. We will also reduce the dataset to different dimensions ranging from 10 to 60 in steps of 10, to see how the model's performance is affected.

About the Dataset

I will briefly describe the dataset so that you will be better equipped to interpret the results. The dataset contains the measurements for the hand gestures. A medical diagnostic procedure called electromyography (EMG) takes these measurements. The dataset has about 11,000 instances, stored across four csv files. The dataset is well balanced and does not contain nulls. Figure 2.11 depicts the data distribution for the four classes that it contains.

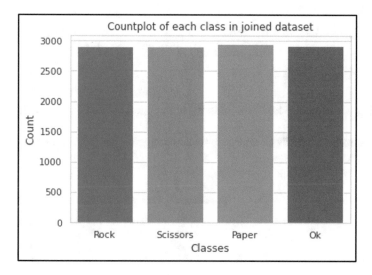

Fig. 2.11 Distribution of classes in hand gestures dataset

Loading Dataset

As they provided the data in four different files, after reading the data, you will need to combine the outputs. On the combined output, you will extract features and the target. Then, you will create training and testing datasets.

You can refer to the project source for these details.

Model Building

As said earlier, I am going to reduce the dataset dimensions from 64 down to various values and then check the model's accuracy on the reduced datasets. For reference, we need to know the model's accuracy on the full dataset. I will use *CatBoost* classifier for model building. We will apply the same algorithm on the reduced datasets and compare their outputs with this reference to the full dataset.

The following code applies the *CatBoost* classifier on our previously prepared dataset:

```
model = CatBoostClassifier(iterations=300,
                           learning_rate=0.7,
                           random_seed=42,
                           depth=5)


model.fit(X_train_scaled, y_train,
          cat_features=None,
          eval_set=(X_test_scaled, y_test),
          verbose=False)
```

After training the model, check the classification report, which is shown in Fig. 2.12.

We will use the above results as reference while comparing the PCA results.

Fig. 2.12 Classification report

	precision	recall	f1-score	support
0	0.98	0.98	0.98	437
1	0.96	0.97	0.97	428
2	0.94	0.95	0.94	449
3	0.93	0.92	0.92	438
accuracy			0.95	1752
macro avg	0.95	0.95	0.95	1752
weighted avg	0.95	0.95	0.95	1752

PCA for Visualization

First, we will try to visualize the effect of features on the target. Later, we will study the effect of reduced features count on the model's accuracy.

As there are 64 columns, we cannot generate a plot of 64 dimensions. So, we first try to reduce the dimensions to two, so that we will create a 2-D plot. You reduce the dimensions from 64 to 2 using the following code:

```
PCA1=PCA(n_components=2)
PCA1=PCA1.fit(X)
```

Now, how do we know if the reduced data represents our original dataset properly? To check upon this, we look at the explained variance ratio with the following command:

```
PCA1.explained_variance_ratio_
```

This is the output:

```
array([0.07220324, 0.07008467])
```

The explained variance ratio is a metric that lets you know the variance percentage for each of the reduced components. In other words, it indicates how close the new components represent the original dataset. The values in the above output are too low and obviously cannot represent our dataset for visualization. We will still apply it on our dataset and study the variance on each data point with the following code:

```
PCA1=PCA1.fit_transform(X)

PCA_df=pd.DataFrame(PCA1, columns = [

            'principal component 1',

            'principal component 2'])

PCA_df
```

Figure 2.13 shows a partial output of variance values for the two components.

Figure 2.14 shows the distribution of four target values for the entire range of principal components plotted on a logarithmic scale.

As you see, the clustering is not so significant. Let us now try to reduce the dataset to three dimensions for visualization. You reduce the dataset to three dimensions by changing the value of the input parameter in *PCA* method call:

```
PCA2=PCA(n_components=3)
PCA2_df=PCA2.fit(X)
```

1 to 25 of 11678 entries   Filter

index	principal component 1	principal component 2
0	-47.34205873851959	-56.87682567108014
1	-127.9429726928701	-130.10187060740384
2	-65.9909968457039	0.6951076174470674
3	-82.93210129634056	-48.28258976200117
4	108.09879507369762	114.75267656902992
5	58.358775109461924	-40.14517032793596

Fig. 2.13 Variance values for two principal components

Fig. 2.14 Two
component PCA

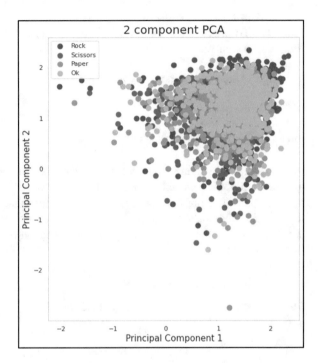

You may check the explained variance ratio as before:

```
PCA2_df.explained_variance_ratio_.sum()
```

The output is:

```
array([0.07220324, 0.07008464, 0.06072772])
```

Again, this certainly is not a good representation of our original dataset. The 3-D
plot generated on the reduced dataset is shown in Fig. 2.15.

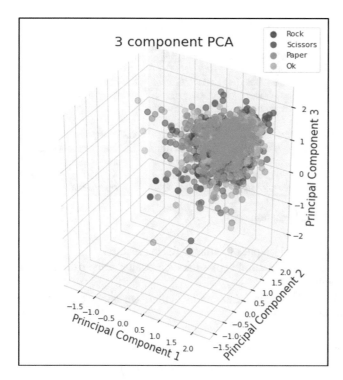

Fig. 2.15 Three component PCA

These 2-D and 3-D visualizations have not helped us much in this case. Apply the technique that you have learned here on your other datasets, and you may get a better visualization on those datasets. Data scientists often use PCA techniques to reduce the datasets to lower dimensions for visualizations.

PCA for Model Building

We will now use the PCA technique for model building. We will reduce our dataset from 64 to 30 dimensions. We use the similar code as earlier with the number of components value set to 30.

```
features = 30

PCA3=PCA(n_components = features)

PCA3=PCA3.fit_transform(X)

PCA3=pd.DataFrame(PCA3)

PCA3
```

index	0	1	2	3	
0	-47.499851910708	-56.9734380180205	-27.878479320096794	-11.408282243156217	116.625
1	-127.90862845163407	-130.07311440336989	-12.03537400483464	7.047532710090076	51.9538
2	-65.91651761640121	0.7460521699641053	161.22107068152852	17.375999345434455	-90.916
3	-82.76573689190444	-48.18584694340971	33.91354850022758	-59.646946922659204	-57.3235
4	108.17067094539301	114.75528299710977	-57.26952319234508	76.25576192583078	-7.6788
5	58.286874204335845	-40.16407961973881	40.34018395129864	-69.50499390986435	-99.673
6	84.11194447872249	17.22232505353061	57.81266962883254	88.99279608592886	26.927
7	34.555683009189735	-96.52165777101058	-132.27692549361012	-82.72018397292987	34.5813
8	71.15277931528856	21.286125191192518	135.46623339192348	-101.51207617387992	-140.935
9	-3.990880623222155	-35.635739565173644	109.25361278687035	6.411453937564321	1.0024
10	-106.79021484880437	-143.4100130936295	41.96982550395796	71.43899924658815	7.6018

Fig. 2.16 Variance values for different features

Figure 2.16 shows the partial output of variance for the features.

As you see, there is a lot of variation across the data points; some exhibit a low variance and some a high.

After preparing the training/testing datasets as earlier, we apply the *CatBoost* classifier on this reduced dataset.

```
model_pca = CatBoostClassifier(iterations=300,
                               learning_rate=0.7,
                               random_seed=42,
                               depth=5)


model_pca.fit(X_train_scaled, label_train,
              cat_features=None,
              eval_set=(X_val_scaled, label_val),
              verbose=False)
```

We check the model's accuracy score:

```
prediction_pca=model_pca.predict(X_val_scaled)
model_pca.score(X_val_scaled, label_val)
```

The output is:

```
0.8333333333333334
```

This is a fairly good accuracy considering that we have reduced the features by more than 50%. The classification report for our new model is shown in Fig. 2.17.

	precision	recall	f1-score	support
0	0.98	0.93	0.95	449
1	0.80	0.86	0.83	444
2	0.79	0.74	0.77	437
3	0.76	0.80	0.78	422
accuracy			0.83	1752
macro avg	0.83	0.83	0.83	1752
weighted avg	0.84	0.83	0.83	1752

Fig. 2.17 Classification report on the revised model

1 to 7 of 7 entries   Filter   ?

level_0	Explained variance
5	0.31086392693212567
10	0.5181312810459301
20	0.7464829448100295
30	0.8676659390603643
40	0.9423504973211887
50	0.9827084554284041
60	0.9982425560206705

Fig. 2.18 Explained variance matrix

Compare this with our classification report on the full dataset, which we can say is comparable and we can easily adapt the reduced dataset for our model.

You may further experiment by reducing the dimensions to various sizes between 5 and 60 as shown here:

```
pca_list = []


for n in [5, 10, 20, 30, 40, 50, 60]:
  PCAmod = PCA(n_components=n)
  PCAmod.fit(X)
  pca_list.append(
       PCAmod.explained_variance_ratio_.sum())
```

You can print the metric of explained variance as follows:

```
pd.DataFrame(pca_list,
       index=[[5, 10, 20, 30, 40, 50, 60]],
       columns=['Explained variance'])
```

Figure 2.18 shows the output matrix.

As you see, for higher values of dimensions, say 30 and above, the variance is acceptable. That to say, the reduced datasets mostly resemble the original one.

You can also check the model's accuracy on the various dataset sizes:

```
accuracies = []


for n in [5, 10, 20, 30, 40, 50, 60]:
 PCAmod = PCA(n_components=n)
 PCAmod = PCAmod.fit_transform(X)
 PCAmod = pd.DataFrame(PCAmod)
 df_pca = pd.concat([PCAmod, pd.DataFrame
            (Y, columns=['label'])], axis = 1)
 X_train, X_val, label_train, label_val =
            train_test_split(df_pca.iloc[:,:-1],
            df_pca.iloc[:,-1], test_size=0.15,
            random_state=42)
 X_train_scaled = s.fit_transform(X_train)
 X_val_scaled = s.transform(X_val)
 model_pca.fit(X_train_scaled, label_train,
            cat_features=None, eval_set=(X_val_scaled,
            label_val), verbose=False)
 accuracies.append(model_pca.score(
            X_val_scaled,label_val))
```

Print the accuracy metric:

```
pd.DataFrame(accuracies,
            index=[[5, 10, 20, 30, 40, 50, 60]],
            columns=['accuracy'])
```

Figure 2.19 shows the output.

As you see, above 30, the model's accuracy score is 80% and above. Even if you consider the reduced dataset of 40 features that gives an accuracy of 86.75%, it is a considerable reduction in dataset size from 64 to 40 columns. Note that the full dataset had given the accuracy score of 95.14%.

The full source for this PCA project is available in the book's repository.

All said, PCA is a powerful technique that is widely used by data scientists for both data visualizations and model building.

level_0	accuracy
	1 to 7 of 7 entries Filter ❓
5	0.4417808219178082
10	0.5907534246575342
20	0.6809360730593608
30	0.8110730593607306
40	0.867579908675799
50	0.898972602739726
60	0.9103881278538812

Fig. 2.19 Accuracy matrix

I will now introduce you to the next technique, and that is independent component analysis.

Independent Component Analysis

Independent component analysis (ICA) is a well-known statistical technique for
detecting hidden factors such as variables, signals, or measurements in a dataset.
Like PCA, it divides multivariate signals into subcomponents. We assume the
components to have non-Gaussian distribution and are independent of each other.
For example, in sound engineering, it is used to determine the sources of sound
signals. Herault and Jutten invented the technique in 1986 for solving the BSS (blind
source separation) problem in signal processing. ICA is more aggressive compared
to PCA in detecting independent subcomponents and is probably more accurate
than PCA.

The sklearn implements ICA in *sklearn.decomposition.FastICA* class. This is the
way it is applied to our dataset:

```
from sklearn.decomposition import FastICA


X = df.drop('Loan_Status', axis=1)
y = df['Loan_Status']


ica = FastICA(n_components=2,
              max_iter=500,
              random_state=1000)
fast_ica=ica.fit_transform(X)
```

After the reductions, the plot generated for the two component analysis is shown
in Fig. 2.20.

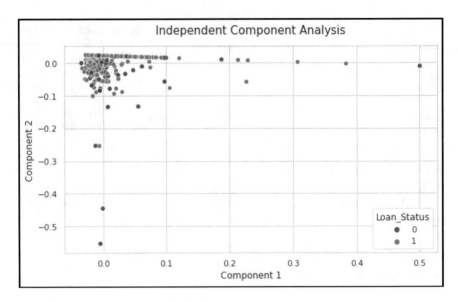

Fig. 2.20 Independent component analysis

Isometric Mapping

This method performs the nonlinear dimensionality reduction through isometric mapping. It extends kernel PCA. We use the geodesic distance to the nearest neighbors for connecting instances. You specify the number of neighbors as a hyper-parameter to *isomap*. The following code snippet shows how to use it.

```
import seaborn as sns
from sklearn.manifold import Isomap


X = df.drop('Loan_Status', axis=1)
y = df['Loan_Status']


isomap = Isomap(n_neighbors=5, n_components=2,
                eigen_solver='auto')
X_isomap = isomap.fit_transform(X)
```

Figure 2.21 shows the isometric chart after component separation.

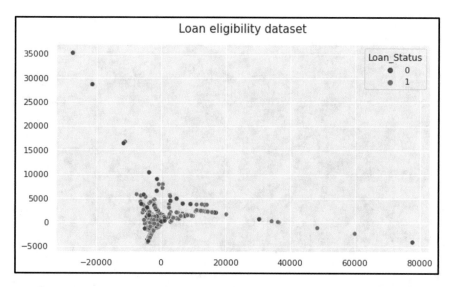

Fig. 2.21 Isometric chart on loan eligibility dataset

t-Distributed Stochastic Neighbor Embedding (t-SNE)

This is another nonlinear dimensionality reduction technique, which is widely used in data visualization, image processing, and NLP. If the number of features is over 50, I recommend that you first use PCA or truncated SVD on your dataset. The following code snippet shows its use:

```
from sklearn.pipeline import Pipeline
from sklearn.manifold import TSNE

sc = StandardScaler()
pca = PCA()
tsne = TSNE()
tsne_after_pca = Pipeline([
    ('std_scaler', sc),
    ('pca', pca),
    ('tsne', tsne)
])
```

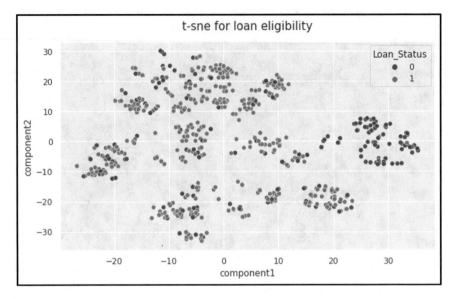

Fig. 2.22 t-sne plot for loan eligibility

```
X = df.drop('Loan_Status', axis=1)
y = df['Loan_Status']


X_tsne = tsne_after_pca.fit_transform(X)
```

The output produced after applying t-SNE transformation is shown in Fig. 2.22.

UMAP

UMAP (Uniform Manifold Approximation and Projection) is a manifold learning technique that competes with t-SNE for visualization quality. During dimension reduction, it preserves more of the global structure as compared to t-SNE and provides a better runtime performance. It is scalable to very large dimensions, making it more suitable for real-world data.

The method uses the concept of *k*-nearest neighbor and stochastic gradient descent (SGD) for optimization. It computes the distances in high-dimensional space, projects them onto the low dimensional space, again computes the distances in this low dimensional space, and finally uses stochastic gradient descent to minimize the differences between these distances.

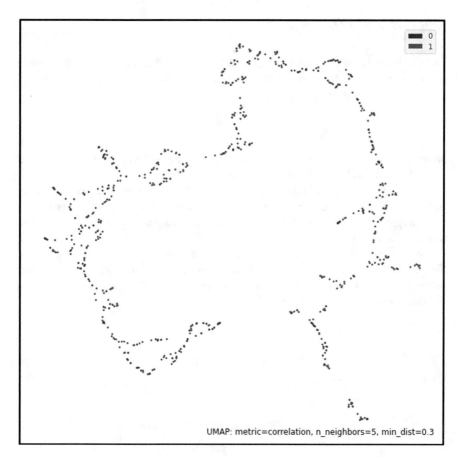

Fig. 2.23 UMAP plot

The sklearn library provides the implementation for UMAP, and thus you may consider this as a drop-in replacement for t-SNE. You create the visualization by calling the UMAP function:

```
mapper = umap.UMAP(n_neighbors=5,
                   min_dist=0.3,
                   metric='correlation').fit(X)
umap.plot.points(mapper, labels=y)
```

The generated plot is shown in Fig. 2.23.

Compare the above output with the correlation map generated by t-SNE and you will realize that UMAP has produced better results.

Singular Value Decomposition

Mathematically, single value decomposition means factorization of any matrix $(m \times n)$ into a unitary matrix U $(m \times m)$, diagonal matrix $(m \times n)$, and a unitary matrix $(m \times n)$. It is expressed as:

$$M = U \, \Sigma V^*$$

where U is an $m \times m$ unitary matrix, Σ is an $m \times n$ diagonal matrix, and V is an $n \times n$ unitary matrix, with V^* being the conjugate transpose of V.

The singular value decomposition (SVD) is a popular method of dimensionality reduction that works far better with sparse data. A typical case of sparse data is collecting the user rating on an e-commerce site. Many times, the user leaves this column blank while providing you the review.

The sklearn library provides its implementation in *sklearn.decompostion. TruncatedSVD* class. It takes just a few arguments, such as number of components, solver algorithm to use, number of iterations, and others. The SVD is applied using the following code:

```
def get_models():
 models = dict()
 for i in range(1,11):
    steps = [('svd',
               TruncatedSVD(n_components=i)),
              ('m', LogisticRegression())]
    models[str(i)] = Pipeline(steps=steps)
 return models
```

To evaluate the model's performance, we define a function as follows that uses cross-validation strategy:

```
def evaluate_model(model, X, y):
 cv = RepeatedStratifiedKFold(n_splits=10,
            n_repeats=3, random_state=1)
 scores = cross_val_score(model, X, y,
            scoring='accuracy', cv=cv, n_jobs=-1,
            error_score='raise')
 return scores
```

We now evaluate all models in the specified range and print the mean and the standard deviation of individual scores:

```
models = get_models()
results, names = list(), list()
for name, model in models.items():
  scores = evaluate_model(model, X, y)
  results.append(scores)
  names.append(name)
  print('>%s %.3f (%.3f)' % (name,
         mean(scores), std(scores)))
```

This is the output:

```
>1 0.688 (0.004)
>2 0.688 (0.006)
>3 0.687 (0.009)
>4 0.685 (0.009)
>5 0.685 (0.009)
>6 0.686 (0.008)
>7 0.686 (0.008)
>8 0.685 (0.018)
>9 0.682 (0.025)
>10 0.803 (0.038)
```

You may also generate the plot of accuracy versus number of components. Such a plot is shown in Fig. 2.24.

As you observe, the accuracy is more when the number of features is high; however, the standard deviation too has increased, which is certainly undesired. For seven components, probably there is a balance with an accuracy around 0.69 and standard deviation relatively low. Likewise, you can use the output of your SVD to decide on the appropriate size for dimensions in your dataset.

Linear Discriminant Analysis (LDA)

Fisher introduced LDA in 1936 for two classes. In 1948, C. R. Rao generalized it for multiple classes. LDA projects data to lower dimensions, yet ensuring that the variability between the classes is maximized and within the class it is reduced.

LDA is used for classification, dimensionality reduction, and visualizations. It is commonly used for feature extraction in pattern classification problems and has

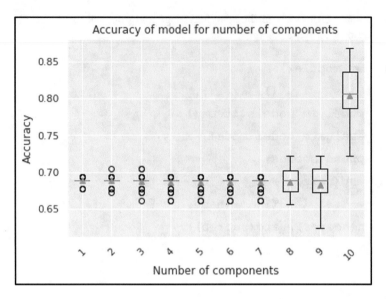

Fig. 2.24 Accuracy vs number of components

outperformed PCA in 2-D and 3-D visualizations. We also use it in computer vision, specifically in face detection algorithms.

The sklearn library provides the LDA implementation in *sklearn. discriminant_analysis.LinearDiscriminantAnalysis* class. I will show its use with a trivial example.

I will use the digits dataset provided in the sklearn library. It has ten components. Using the LDA algorithm, we will reduce this number between 2 and 9. We apply the algorithm using the following statement:

```
lda = LinearDiscriminantAnalysis(n_components=comps)
lda.fit(X_train, y_train)
```

The algorithm will fit our training dataset into the specified number of components. We will then apply the random forest classifier on the transformed dataset using the following code snippet:

```
classifier = RandomForestClassifier(
                  random_state=1, max_depth=10)
classifier.fit(lda.transform(X_train), y_train)
```

We will check the accuracy score on the test dataset using the statement:

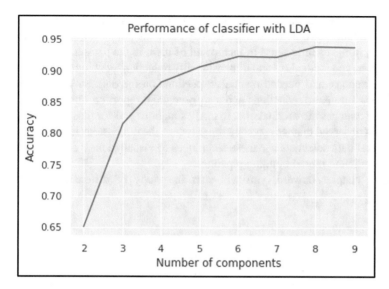

Fig. 2.25 LDA performance plot

```
y_pred = classifier.predict(lda.transform(X_test))
accuracies.append(
            metrics.accuracy_score(y_test,y_pred))
```

Finally, we will plot the accuracy for the number of components that we tested in the range 2–9. The classifier performance plot is shown in Fig. 2.25.

From the plot, we clearly observe that the accuracy increases as we add more features until it reaches a value of about 8. The accuracy for 8 and 10 components (the full set) almost remains the same. Thus, we have achieved the feature reduction from 10 to 8. Looking at the chart, where we observe the accuracy of about 0.90 for component count of 5, we may decide on the features reduction to 5. Did you see the power of the LDA algorithm? For datasets with very high dimensions, this is a powerful technique for reducing the dataset dimensions.

The full source code for this project is available in the book's repository.

Summary

Though having many features for analysis and machine learning model building is an advantage, we also consider it a serious drawback. We also call it the curse of dimensionality. Larger count of dimensions makes the data visualizations difficult. We human beings can probably understand only 2-D and 3-D plots. The large number of dimensions also increases the resource and time requirements for model building. Having larger dimensions does not always mean that we will build better models. Many-a-times, two or more features may carry the similar information. In such situations, it makes little sense to include all of them in the dataset.

In this chapter, you studied several dimensionality reduction techniques. Some of these were elementary and needed manual inspections on the dataset. Techniques such as PCA do automatic reductions and also generate a report on how close the reduced dataset matches the original one. In many situations and for those advanced techniques, you observed that we can sometimes reduce the dimensions to even 50% or better. Several data scientists use these techniques to visualize their datasets and creating more efficient model building pipelines.

In the next chapter onward, you will start the study of different classical algorithms.

Part I
Classical Algorithms: Overview

In machine learning, the regression and classification are the two major tasks. We can solve both using statistical techniques. In regression, you attempt to determine the strength and character of a relationship between a dependent variable and a set of independent variables. In classification, you arrange your data in groups or categories according to some established criteria. Determining the price of a house based on its features is a regression problem. The features here could be like floor size, number of bedrooms, its location, and so on. The target is the price of the house. Determining if the breast cancer is malignant or benign depending on certain clinical measurements is a binary classification problem. Recognizing handwritten digits as numeric 0 through 9 is a multi-class classification problem.

Statisticians and mathematicians have developed several algorithms for both regression and classification. We use a labeled dataset for training these algorithms. Thus, they fall under the category of supervised learning. You will study the clustering algorithms in the latter part of this book, which are based on unsupervised learning. Researchers have developed a few algorithms only for regression, while we can use most others for both regression and classification.

The regression is a well-studied statistical technique for predictive analysis. Statisticians have done a lot of work on regression, and we have solutions in every domain to solve regression tasks. There are several types of regression—linear, polynomial, ridge, lasso, elastic net, Bayesian linear, logistic, and so on. I have devoted a full chapter on regression.

The other machine learning algorithms are decision trees, K-nearest neighbors (KNN), naive Bayes, support vector machines (SVM), and so on. These algorithms can solve both regression and classification problems. The decision tree is one of the early algorithms in this category and learners easily understand it because of its similarity to the human thinking process. As decision trees can be extremely large for huge datasets, we apply the statistical ensemble techniques, like bagging and boosting, to improve its performance.

You will begin your study of classical algorithms with regression and then move on to other algorithms like decision trees, KNN, naive Bayes, and SVM. You will also learn the various bagging and boosting techniques.

So, we begin with regression.

Chapter 3
Regression Analysis

A Well-Studied Statistical Technique for Predictive Analysis

In a Nutshell

Regression is one of the well-understood and studied algorithms in statistics. It has been successfully applied to machine learning and has been extensively used in many practical applications. This is a predictive modeling technique that establishes the relationship between a dependent and one or more independent variables. There are many forms of regression—linear and logistic regression are the most widely used among those. Regression has a wide variety of applications in forecasting and time series modeling. It is also used for establishing the causal effect relationship between many variables.

When to Use?

Regression analysis has been used successfully in a wide range of applications. Whenever you want to establish a relationship between several independent variables to predict a target value, you use regression. As an example, consider estimating how much your house in Boston would fetch if put on sale. The price of the house is decided by several independent variables such as its location, the number of bedrooms it has, the floor size, and several other factors. These independent variables are called *predictors* or *features* in machine learning.

Businesses make a frequent use of regression analysis for forecasting opportunities and threats. We use it for demand analysis, predicting the highest bid, forecasting the number of consumers on a festival day, forecasting the insurance claims in the coming year, and so on. The financial analysts use it to forecast corporate returns. We also use it in weather forecasting and in predicting stock prices. The applications are many.

Let us now look at the various kinds of regression.

© The Author(s), under exclusive license to Springer Nature Switzerland AG 2023
P. Sarang, *Thinking Data Science*, The Springer Series in Applied Machine Learning,
https://doi.org/10.1007/978-3-031-02363-7_3

Regression Types

The first algorithms that students learn in data science are the linear and logistic regressions. Because of their popularity, most think that these are the only two regression types. There are innumerable forms of regressions, each having a specific purpose for which it was designed. The regression types are mostly driven by three metrics:

- Number of dependent variables
- Type of dependent variables
- Shape or the hyperplane of regression boundary

   I will discuss a few of the most popular regressions. These are listed here:

- Linear
- Polynomial
- Ridge
- Lasso
- EasticNet
- Logistic

   I will describe the working of each one and its purpose. Toward the end, I will provide you with a few guidelines on how to select the regression model for your business use case.

Linear Regression

This is probably the first type of rigorously studied regression analysis and extensively used. The linear regression assumes a linear relationship between the target and its predictors. Thus, it is easier to fit the model on such datasets, and the resulting estimators can be easily validated. In linear regression, you find the best-fit straight line that establishes a relationship between a dependent variable (y) and one or more independent variables (x). The independent variables may be continuous or discrete. You need to find the best fit line, which is also called the regression line. For any unseen x, you can now determine the corresponding y by mathematically solving the equation of the regression line. Mathematically, the equation of the regression line is specified as:

$$y_i = \beta_0 + \beta_1 x_{i1} + \cdots + \beta_p x_p + \varepsilon_i = x_i^T \beta + \varepsilon_i, \qquad i = 1, \cdots, n$$

where ^{T} denotes the transpose, so that $x_i^T \beta$ is the inner product between vectors x_i and β.

   ε is the noise term in the linear relationship between the dependent variable and regressors.

Generally, these n equations are written in matrix notation as:

$$y = X\beta + \varepsilon$$

where:

$$y = \begin{pmatrix} y_1 \\ y_2 \\ \vdots \\ y_n \end{pmatrix}$$

$$X = \begin{pmatrix} x_1^T \\ x_2^T \\ \vdots \\ x_n^T \end{pmatrix} = \begin{pmatrix} 1 & x_{11} & \cdots & x_{1p} \\ 1 & x_{21} & \cdots & x_{2p} \\ \vdots & \vdots & \ddots & \vdots \\ 1 & x_{n1} & \cdots & x_{np} \end{pmatrix}$$

$$\beta = \begin{pmatrix} \beta_0 \\ \beta_1 \\ \beta_2 \\ \vdots \\ \beta_p \end{pmatrix}$$

$$\varepsilon = \begin{pmatrix} \varepsilon_1 \\ \varepsilon_2 \\ \vdots \\ \varepsilon_n \end{pmatrix}$$

You use the least square method to get the best fit line. In this method, you minimize the sum of the squares of the vertical deviations from each observed data point to the regression line. Mathematically, it is expressed as:

$$\min_{\beta} \| X\beta - y \|_2^2$$

As seen in Fig. 3.1, we assume the observations to be random deviations from the underlying relationship between y and x.

Though the linear regression has given us excellent results, it is still based on a few assumptions that I will discuss now.

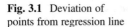

Fig. 3.1 Deviation of
points from regression line

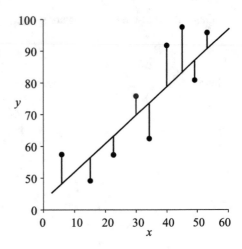

Assumptions

In the list squared estimation technique, we make certain assumptions about the
predictor variables, target, and their relationship. Statisticians have extended this
standard estimation technique with many extensions that I will discuss later. Though
these extensions make the estimation process more complex and time-consuming,
they reduce the assumptions to a weaker form and, sometimes, eliminate them in
their entirety. So, let us look at the assumptions.

We assume that there is no error in measurement of the predictor variables x,
which is an unrealistic assumption. If there is an error in measurement of predictor
variables, we assume that the variance in errors is independent of their values. We
call this *homoscedasticity* the assumption of constant variance. This is usually not
true. The standard deviation on the predictor values is proportional to their mean,
rather than being a constant. Another assumption that we make is that the errors in
the response variables are uncorrelated with each other. The multicollinearity in
predictor variables can seriously affect the parameter estimation. Most importantly,
we make a linearity assumption such that the mean of a response variable is a linear
function of the regression coefficients and the predictor variables.

From these assumptions, what I want to point out here is that you cannot simply
apply the linear regression to any dataset for predictive analysis. A thorough analysis
of the dataset (EDA) is required, and some transformations, scaling on data points,
may be necessary before applying the linear regression. After all, the relationship
may be polynomial or could be some complex hyperplane. During exploratory
analysis, if you observe such trends, apply other regression techniques that are
discussed further.

Polynomial Regression

In this type of regression, we model the relationship between the independent variable x and the dependent variable y as the nth degree polynomial in x. Mathematically, we express it as:

$$y_i = \beta_0 + \beta_1 x_i + \beta_2 x_i^2 + \cdots + \beta_m x_i^m + \varepsilon_i, \quad (i = 1, 2, \cdots, n)$$

In matrix notation, the entire model is written as:

$$\vec{y} = X\vec{\beta} + \vec{\varepsilon}$$

From these equations, you can understand that the polynomial regression is technically a special case of multiple linear regression. The best fit is found by treating each x_i as an independent variable and applying the linear regression on each. The combination of these multiple regressions results in a final polynomial. Figure 3.2 shows a polynomial regression model.

While fitting the polynomial on your dataset, you must observe the degree of your polynomial as it may cause either underfitting or over-fitting, as depicted in Fig. 3.3.

Fig. 3.2 Polynomial regression

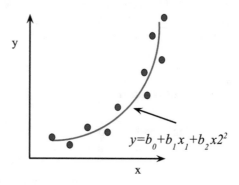

$$y = b_0 + b_1 x_1 + b_2 x2^2$$

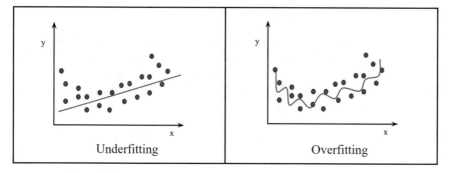

Underfitting　　　　Overfitting

Fig. 3.3 Underfitting/over-fitting of regression line

As with linear regression, the outliers can seriously affect the values of coefficients.

I will now discuss the regression models that use some statistical regularization techniques to overcome the assumptions that we have discussed so far.

Ridge Regression

When you observe multicollinearity (independent variables are highly correlated) in your dataset, you will use ridge regression. Ridge is a special case of Tikhonov regularization. It regularizes all parameters equally. Generally, multicollinearity is observed in datasets having many features. It improves the parameter estimates at the cost of a tolerable amount of bias. It adds a certain degree of bias to the regression estimates to reduce standard errors. Mathematically, it is expressed as:

$$\underset{\beta \in R^p}{\operatorname{argmin}} \underbrace{\|y - X\beta\|_2^2}_{\text{Loss}} + \underbrace{\lambda\|\beta\|_2^2}_{\text{Penalty}}$$

where the first term is our least square term and the second one is the Lagrange multiplier of the constraint. When λ becomes zero, the estimator reduces to ordinary least squares. For all other values, it shrinks the least square term to have a low variance. This is the *l2* regularization.

Lasso Regression

As in ridge, lasso too penalizes the absolute size of the regression coefficients. Mathematically, it is expressed as:

$$\underset{\beta \in R^p}{\operatorname{argmin}} \underbrace{\|y - x\beta\|_2^2}_{\text{Loss}} + \underbrace{\lambda\|\beta\|_1}_{\text{Penalty}}$$

The penalty term is now the absolute value instead of the squares. A large penalty would shrink the estimate toward absolute zero, resulting in the elimination of the variable. This helps us in feature selection. Lasso is a *l1* regularization method. When you have a group of predictors that are highly correlated, lasso picks up one of them and shrinks the others to zero.

ElasticNet Regression

This is a hybrid of lasso and ridge regressions. Mathematically it is expressed as:

$$\widehat{\beta} = \underset{\beta}{\operatorname{argmin}} \left(\| \, y - X\beta \|^2 + \lambda_2 \, \| \, \beta \|^2 + \lambda_1 \, \| \, \beta \|_1 \right)$$

As you can see, it trained the model with both *l1* and *l2* regularizers. If your dataset has multiple predictors which are correlated, elastic-net will probably pick up all, while lasso will pick up one of these at random. Thus, it encourages a group effect on highly correlated predictors, and there is no limitation on the number of selected predictors. The disadvantage is that it can suffer from double shrinkage.

I will now discuss sklearn's implementation of linear regression models.

Linear Regression Implementations

The sklearn library provides ready-to-use implementations for linear, ridge, and lasso regressions. To demonstrate the use of these classes, I have created a small application to do a regression fit on the insurance dataset taken from Kaggle. The entire project source code is available in the book's download site. I will use some code snippets here to show you how to use the built-in regression classes.

Linear Regression

The sklearn library implements the linear regression algorithm in *sklearn. linear_model.LinearRegression* class. You call it using the following statements:

```
from sklearn.linear_model import LinearRegression

linearRegression = LinearRegres-
sion().fit(X_train_scaled, label_train)
```

After the algorithm fits the regression line on the training dataset, you use it for prediction on the test or the unseen data using the following statement:

```
y_pred = linearRegression.predict(X_val_scaled)
```

You print the error metrics using the following statement:

```
error_metrics(y_pred,label_val)
```

It will print MSE, RMSE, and coefficient of determination as follows:

```
MSE: 0.1725732138508936
RMSE: 0.4154193229146829
Coefficient of determination: 0.7672600627397954
```

The following statement prints the model score as shown:

```
linearRegression.score(X_val_scaled, label_val)
```

```
0.8073560218213742
```

Ridge Regression

The sklearn implements the ridge regression with built-in cross-validation in *sklearn. linear_model.RidgeCV* class. You apply this model on your dataset using the following statements:

```
from sklearn.linear_model import RidgeCV
```

```
alphas = [0.007, 0.009, 0.002, 0.01]
```

```
ridgeCV = RidgeCV(alphas=alphas, cv=4).fit(
                   X_train_scaled, label_train)
```

You do the prediction using the fitted model like this:

```
ridgeCV_pre = ridgeCV.predict(X_val_scaled)
```

You can print the alpha value and the error metrics as follows:

```
print('Alpha found: ',ridgeCV.alpha_)
error_metrics(ridgeCV_pre,label_val)
```

```
Alpha found:  0.01
MSE:  0.17256828552086406
RMSE:  0.4154133911188517
Coefficient of determination:  0.7672565992373196
```

As before, you print the model's score as follows:

```
ridgeCV.score(X_val_scaled, label_val)
```

```
0.8073615233304525
```

Lasso Regression

The sklearn implements lasso regression with cross-validation in.

```
sklearn.linear_model import LassoCV
```

You use the model using following code snippet:

```
from sklearn.linear_model import LassoCV

alphas2 = np.array([1e-6, 5e-6, 1e-5, 2e-5, 1e-3])

lassoCV = LassoCV(alphas=alphas2,
                  max_iter=5e4, cv=3).fit(
                  X_train_scaled, label_train)
```

You do the predictions and print the error metrics as before:

```
lassoCV_pre = lassoCV.predict(X_val_scaled)

print('Alpha found: ',lassoCV.alpha_)
error_metrics(lassoCV_pre,label_val)
```

```
Alpha found:  2e-05
MSE:  0.17253931207747167
RMSE:  0.415378516629678
Coefficient of determination:  0.7672545446512832
```

Print the model score like this:

```
lassoCV.score(X_val_scaled, label_val)
```

```
0.8073938664691824
```

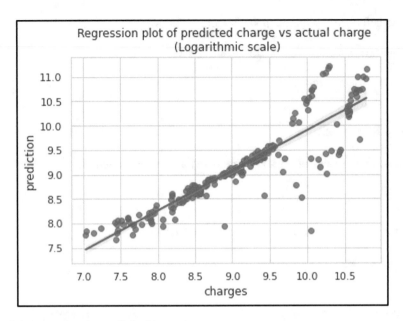

Fig. 3.4 Regression plot prediction/charges

The three outputs show that all three models give out almost identical scores with lasso being the highest.

We show the plot of prediction versus actual charges along with the regression line in Fig. 3.4.

Do look up the entire project source as it contains lots of EDA and features engineering.

Bayesian Linear Regression

So far, you have learned simple linear regression with the effects of regularizations. Will these models work well if you have insufficient data or poorly distributed data? This is where Bayesian linear regression (BLR) comes to your help. The BLR not only helps in finding the single best value for the model parameters but also helps in determining the posterior distribution for the model parameters. We base BLR on the famous Bayes theorem. Bayesian statistics are a probability-based mathematical method for solving statistical problems. It helps in creating beliefs in the evidence of the new data. Thus, this technique is ideally suited if you do not have enough data points in your dataset.

Bayes' theorem states as shown in Fig. 3.5.

Fig. 3.5 Posterior
probability in Bayes'
theorem

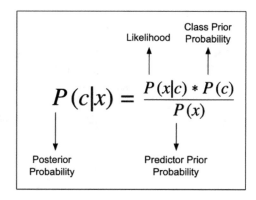

where:

- $P(c|x)$ is the posterior probability of class c given predictor x
- $P(c)$ is the prior probability of class c
- $P(x|c)$ is the probability (likelihood) of a predictor given class c
- $P(x)$ is the prior probability of predictor x

Applying this theorem on your dataset, it allows you to put a prior on both the coefficients and the noise; you use these priors in the absence of data. More importantly, BLR tells you about which parts of the data it is confident about and which parts are based entirely on the priors. Specifically, you can investigate upon the following:

- What is its confidence on the estimated relation?
- What is the estimated noise and the posterior distribution on it?
- What is the estimated gradient and the posterior distribution on it?

Without going into the mathematics of BLR, let us look at its implementation.

BLR Implementation

The sklearn implements BLR in *sklearn.linear_model.BayesianRidge* class. The instantiation requires several parameters; however, the defaults are provided for all of them, so you do not have to really worry about those. Only, if you want to do some fine-tuning, you will have to fiddle with them.

I will now discuss a small project that shows the BLR use in a practical situation.

BLR Project

The project uses a dataset on combined-cycle power plant published in UCI repository. The dataset contains about 9000 data points. The plant's output (PE) depends on four features—temperature (AT), exhaust vacuum (V), ambient pressure (AP), and relative humidity (RH).

After reading the dataset, you will set up the features and the target using the statements:

```
features=['AT','V','AP','RH']
target='PE'
```

Set some parameter values:

```
Alpha1=0.00006138969568741323
Alpha2=12267443.02717635
Lambda1=5.936316029785909e-10
Lambda2=3075473.158359934
```

And then apply BLR on the training dataset as follows:

```
model=BayesianRidge(alpha_1=Alpha1,
                    alpha_2=Alpha2,
                    lambda_1=Lambda1,
                    lambda_2=Lambda2)
model.fit(x_train,y_train)
```

After the training, check the model's accuracy on the test data:

```
y_pred=model.predict(x_test)
print("Accuracy score
            {:.2f} %\n".format(
            model.score(x_test,y_test)*100))


Accuracy score 92.35 %
```

You may also print the R2, MAE, and MSE as follows:

```
print("R2 Score: {:.2f} %".format(
            r2_score(y_test,y_pred)*100))
print("Mean Absolute Error {:.2f}".format(
            mean_absolute_error(y_test,y_pred)))
print("Mean Squared Error {:.2f}".format(
            mean_squared_error(y_test,y_pred)))


R2 Score: 92.35 %
Mean Absolute Error 3.74
Mean Squared Error 22.13
```

Figure 3.6 shows a plot of predicted versus actual values.

The full project source code is available in the book's repository.

With these different types of regression for finding out the best fit regression line, there is another important type of regression that I have not discussed so far and that is logistic regression.

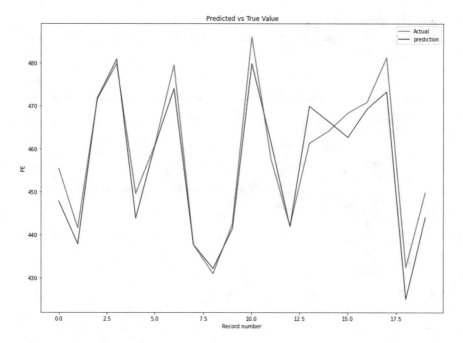

Fig. 3.6 Plot of predicted vs actual

Logistic Regression

Unlike linear regression, where the predictions are continuous, in logistic regression, those are discrete. It will output a binary value such as pass/fail, true/false, yes/no, and so on. When the output is binary, we call it simply binary logistic regression. The values of dependent variables can be ordinal, such as low, medium, and high. Here, we call it ordinal logistic regression. If the dependent variable is multi-class such as cat, dog, lion, and goat, we call it multinomial logistic regression. In all cases, the output is discrete. Thus, logistic regression is considered a classification algorithm. To cite a few practical uses of logistic regression, consider the classification problems, such as email, spam or not spam; tumor, malignant or benign; and online transaction, fraud or non-fraud.

Basically, logistic regression is achieved by passing the output of the linear regression through an activation function that maps the real number to either 0 or 1. This can be seen in Fig. 3.7.

The sigmoid function shown in Fig. 3.8 is used for the activation.

The function value lies between 0 and 1. If the probability is greater than 0.5, we treat the output as 1 else 0.

Mathematically, the logistic regression equation is written as:

$$P = \frac{1}{1 + e^{-(a+bX)}}$$

where P is the probability estimate.

I will now show you the sklearn's implementation of logistic regression model.

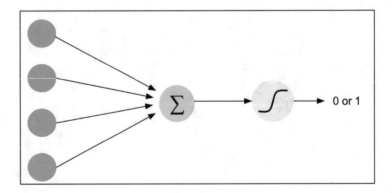

Fig. 3.7 Sigmoid activation in logistic regression

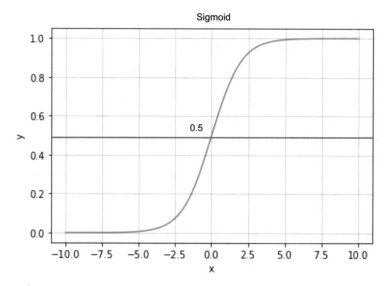

Fig. 3.8 Sigmoid function

Logistic Regression Implementation

Like linear regression, sklearn library provides ready-to-use implementation for logistic regression. To show its use, I have created a small application to do a regression fit for the employee dataset taken from Kaggle. The entire project source code is available on the book's download site. Like before, I will use some code snippets here to show you how to use the logistic regression class.

You apply the logistic regression on your dataset using the following code snippet:

```
from sklearn.linear_model import LogisticRegression

lr = LogisticRegression(solver='liblinear').fit(
                    X_train_scaled, label_train)
```

You apply *l1* regularization as follows:

```
lr_l1 = LogisticRegressionCV(Cs=15, cv=4, penalty='l1',
                    solver='liblinear').fit(
                    X_train_scaled, label_train)
```

And *l2* regularization as in the following statement:

```
lr_l2 = LogisticRegressionCV(Cs=15, cv=4, penalty='l2',
                    solver='liblinear').fit(
                    X_train_scaled, label_train)
```

You print the classification report using the library *supplied sklearn.metrics. classification_report* class. This is illustrated in the code snippet below:

```
from sklearn.metrics import classification_report
print('Classification report for Logistic
            regression without regularization:')
print(classification_report(label_val,y_pred['lr']))

print('Classification report for Logistic regression
            with L1(Lasso) regularization:')
print(classification_report(label_val,y_pred['l1']))

print('Classification report for Logistic regression
            with L2(Ridge) regularization:')
print(classification_report(label_val,y_pred['l2']))
```

The output of the above code snippet is shown in Fig. 3.9.

You print the summary by concatenating the above metrics, which is seen in Fig. 3.10.

You can also print the confusion matrix by using the built-in classes as follows:

```
from sklearn.metrics import confusion_matrix,
                            ConfusionMatrixDisplay

f, (ax1, ax2, ax3) = plt.subplots(1,3,figsize=(18,5))

disp1 = ConfusionMatrixDisplay(
                        confusion_matrix=cm['lr'],
                        display_labels=lr.classes_)
disp1.plot(cmap='Blues', ax=ax1)
ax1.set_title('Logistic')
```

```
⸴ Classification report for Logistic regression without regularization:
              precision    recall  f1-score   support

           0       0.75      0.91      0.82       610
           1       0.71      0.44      0.54       321

    accuracy                           0.75       931
   macro avg       0.73      0.67      0.68       931
weighted avg       0.74      0.75      0.73       931

Classification report for Logistic regression with L1(Lasso) regularization:
              precision    recall  f1-score   support

           0       0.75      0.91      0.82       610
           1       0.71      0.44      0.54       321

    accuracy                           0.74       931
   macro avg       0.73      0.67      0.68       931
weighted avg       0.74      0.74      0.73       931

Classification report for Logistic regression with L2(Ridge) regularization:
              precision    recall  f1-score   support

           0       0.75      0.91      0.82       610
           1       0.71      0.43      0.54       321

    accuracy                           0.74       931
   macro avg       0.73      0.67      0.68       931
weighted avg       0.74      0.74      0.72       931
```

Fig. 3.9 Assorted classification reports

Fig. 3.10 Summary of classification reports

	lr	l1	l2
precision	0.739845	0.738668	0.737487
recall	0.745435	0.744361	0.743287
fscore	0.726927	0.725556	0.724182
accuracy	0.745435	0.744361	0.743287
auc	0.672905	0.671347	0.669790

The generated confusion matrix can be seen in Fig. 3.11.

The complete project source is available in the book's repository.

With the various regression techniques discussed, I will now provide you some guidelines on which one to use.

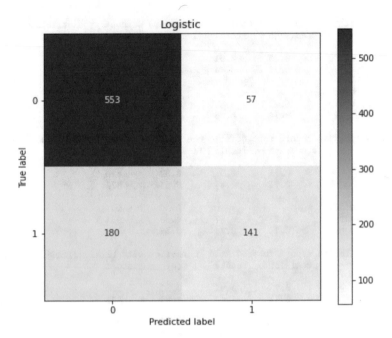

Fig. 3.11 Confusion matrix of logistic regression

Guidelines for Model Selection

From the above discussions, you can easily gather that exploratory data analysis is an important step before narrowing down your selection of the regression model. The model selection, as you have seen, depends on the relationships between the target and its predictor variables. It also highly depends on the correlation between the predictors and the dimensionality of the data. Most of the time, the EDA is a very lengthy process. Personally, I prefer applying different models on the dataset without doing a detailed EDA. Of course, I ensure the data is clean by doing preprocessing wherever required. I then use evaluation metrics to choose the best performing model. If a certain model is performing appreciably better than others, I further work on it to improve its accuracy. We can improve the accuracy by fine-tuning the model parameters and, if required, going back to EDA for refining the dataset.

What's Next?

Besides the above-discussed regression models, there are many more such as Robust, Quantile, and so on. If you look up Wikipedia, you will probably find maybe 20–30 more of them. If you think of studying each one, you will need to take up a full statistics course. As a data scientist, you need not worry about studying all those. I discussed only a few, which are mostly used by data scientists. Implementation of these models is readily available in the popular machine learning libraries. Just learn how to apply those on your datasets. Compare the evaluation metrics and decide which one best fits your purpose.

Summary

The regression is a deeply studied statistical technique that is now applied to machine learning. There are many success stories of its application in practical cases. Though many think that linear and logistic regressions are the only two types of regression, there are many more regression models. They developed these to account for high dimensionality and multicollinearity of datasets. In this chapter, you studied the purpose of a few of them and how they help in handling different datasets. Thus, the selection of the algorithm for your dataset depends heavily on the results of your exploratory data analysis. The scikit-learn and several other libraries provide the implementation of many regression algorithms. As a data scientist, experiment with these pre-implemented algorithms in model development. Select the model that performs the best. If you are not happy with the model's accuracy, try fine-tuning its parameters, and if this still does not work out, go back to EDA. This practically saves you a lot of model development time.

In the next chapter, you will study another important machine learning algorithm and that is the decision tree.

Chapter 4
Decision Tree

A Supervised Learning Algorithm for Classification

In this chapter, you will learn the use of another important machine learning algorithm that is called decision tree.

In a Nutshell

This is a supervised learning algorithm mainly used for classification, but can also be used for regression analysis. Decision trees mimic human thinking while making decisions and thus usually are easy to understand. We can depict the entire tree as a visual representation for further understanding.

To motivate the study of decision trees, first I will discuss the areas and domains where decision trees have been successfully used.

Wide Range of Applications

These are several domains and use-cases where decision trees have been success-fully applied. We use it in medical diagnosis of diseases, fraud detection, customer retention, marketing, educational sector, and so on. In the medical field, physicians and medical practitioners use decision trees in early identification of preventable conditions, such as diabetes or dementia. The banks use decision trees to identify fraudulent behavior of previous customers to prevent fraud, saving them huge amounts of losses. Companies use decision trees for customer retention by analyzing their purchases and releasing new offers based on their buying habits. They may also use decision trees for evaluating the customer satisfaction levels. Businesses observe the competitors' performance on their products and services and build decision trees for customer segmentation, improving the accuracy of their new promotional cam-paigns. The universities can shortlist students based on their merits, attendance,

P. Sarang, *Thinking Data Science*, The Springer Series in Applied Machine Learning,
https://doi.org/10.1007/978-3-031-02363-7_4

grades, etc. by building decision trees. Thus, there are plenty of applications of decision trees.

We used decision trees in the past for several such applications because they mimic human thinking ability while making a decision and are also easily interpretable because of its tree-like structure.

I will now discuss the workings of a decision tree.

Decision Tree Workings

In the use of the decision tree algorithm, it involved two processes:

- Building the tree itself
- Traversing the tree to infer an unseen data point

Out of the above two, building the tree is more complicated, while using the tree for inference is relatively easy. I will first discuss the traversal and then show you how to build a tree by presenting the algorithm for tree building.

Tree Traversal

Understanding tree traversal is very easy as this is the way we humans think while solving any problem. Consider a situation where you would like to know if a player would play tennis given certain weather conditions, such as temperature, humidity, wind, and general weather outlook. In other words, your dataset has four features: wind, humidity, temperature, and outlook. Depending on these four parameters, the target condition "Play Tennis" is either set to yes or no. Assume that somebody has constructed a decision tree to determine the "Play Tennis" status as shown in Fig. 4.1.

Consider a situation where you would like to determine if a player would play tennis under following conditions (feature values):

- Outlook: sunny
- Temperature: mild
- Humidity: normal
- Wind: strong

In other words, we want to find out whether the player will play tennis today when the general weather outlook is sunny and the temperatures are mild with normal humidity and strong winds.

To traverse the tree, we start at its top. We check if the outlook is sunny, rainy, or overcast. As the outlook today is sunny, we take the left branch of the tree and move on to the next decision node that checks the humidity. As the humidity is normal, we

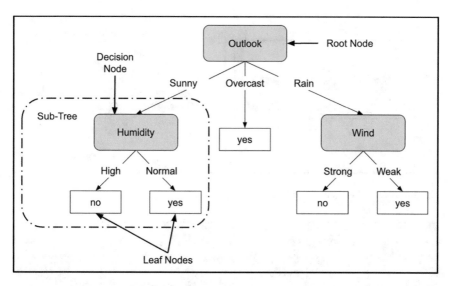

Fig. 4.1 Decision tree to determine whether to play tennis?

take the right branch and land into the leaf node that says yes. So, the conclusion is that the player will play tennis today.

The general procedure is that you start at the top and keep moving down left or right until you hit a leaf node; this is the prediction of your tree search.

Now comes the most important part, and that is how to construct the tree.

Tree Construction

The major question that would come to your point while constructing the tree is: What is the starting point? In other words, who becomes the root node? We have thousands of data points in our dataset, and each one is eligible to be a root. However, selecting an inappropriate data point as a root can easily result in an imbalanced tree. Figure 4.2 shows both balanced and imbalanced trees.

So, it is important to select an appropriate root node. For this, you will need to understand a few terms. I will define those terms first before explaining the tree construction process, that is, the algorithm.

Entropy

Entropy is a measure of how disordered your data is. High entropy means the data is in chaos and low entropy means that the data is uniformly distributed. Consider our previous example, where we have the labeled data, a player is "playing" or "not

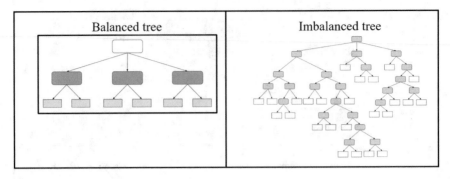

Fig. 4.2 Balanced and imbalanced trees

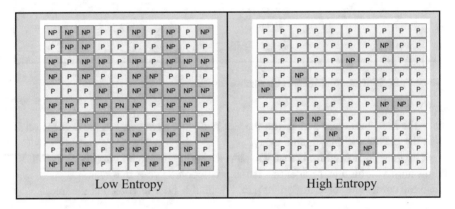

Fig. 4.3 Low and high entropies

playing." Let us call these values "P" for playing and "NP" for non-playing. Figure 4.3 shows both cases of low and high entropies.

Entropy is almost zero when the dataset approaches homogeneity and equals one when it is equally distributed. During tree construction, having low entropy helps in better segregation of classes and thus creates a better tree for predictions.

Mathematically, entropy is expressed as:

$$\text{Entropy}(S) = -(P_\oplus \log_2 P_\oplus + P_\ominus \log_2 P_\ominus)$$

where $P_\oplus$ is the portion of positive examples and $P_\ominus$ is the portion of negative examples in S.

I will show you how this formula is applied when I explain the tree construction. I will now define another term called information gain.

Information Gain

During tree construction, you use information gain for generalizing the entropy. As said earlier, having low entropy helps in creating better predicting models. Mathematically, information gain is defined as follows:

$$\text{Gain}(S, A) = \text{Entropy}(S) - \sum_{v \in \text{Values}(A)} \frac{|S_v|}{|S|} \text{Entropy}(S_v)$$

where *Values(A)* is the all possible values for attribute A and S_v is the subset of S for which attribute A has value v.

At every split of a node, we compute information gain—the higher the value for this will be a better split as high information gain shows low entropy. Low entropy means more even distribution. The even distribution means better segregation into classes. Thus, the information gain can be a good measure to decide which of the attributes are the most relevant. A most relevant attribute is used as a node and split further into branches.

A feature like the credit card number takes many distinct values, resulting in high values of information gain. This can be a drawback while working with datasets having certain columns with many distinct values. Certainly, you will exclude credit card numbers in your analysis, as you do not want to generalize customers based on their credit card numbers.

I will now introduce you to one more term—Gini.

Gini Index

The Gini index is mathematically expressed as:

$$G = \sum_{i=1}^{C} p(i) * (1 - p(i))$$

where C is the number of classes and $p(i)$ is the probability of picking the data point having class i. We may simplify the expression to:

$$G = \sum_{i=1}^{C} p(i) * (1 - p(i))$$

$$= \sum_{i=1}^{C} p(i) - \sum_{i=1}^{C} p(i)^2$$

$$= 1 - \sum_{i=1}^{C} p(i)^2$$

where $p(i)$ is the probability of class i in the attribute. You know that the first term in the above equation that denotes the sum of all probabilities equals 1.

Thus, the Gini index is computed by subtracting the sum of the squared probabilities of each class from the numeric one.

$$\text{Gini}(\text{attribute}_v) = 1 - \sum_{i=1}^{N} (P_i)^2$$

$$\text{Gini}(\text{attribute}) = \sum_{v=\text{values}} P_v * \text{Gini}(\text{attribute}_v)$$

where P_v is the probability of values in attribute and P_i is the probability of picking the data point with the class i.

It is simple to implement and favors splitting into larger partitions. The Gini index varies between 0 and 1, where 0 represents purity of the classification and 1 denotes random distribution of elements among various classes. A Gini index of 0.5 shows that there is equal distribution of elements across some classes. To state it differently, it essentially gives us the probability of incorrectly classifying a selected feature.

I will now come to the important point of tree construction. We have two different implementations for tree construction, one based on information gain and the other that uses Gini index. I will describe the tree construction example based on the IG, which is used in the popular ID3 algorithm.

Constructing Tree

To illustrate the tree construction process, I have created a small dataset. This is a labeled dataset that decides whether a certain player will play tennis based on the current climate. The measure for climate is temperature, humidity, wind, and general outlook. These are the features that will decide the target value—player playing tennis or not. Table 4.1 shows the sample data used for constructing the tree.

I will now show you how to construct a tree for this partial data.

So, let us do the mathematics behind building this tree.

First, we calculate entropy for our complete problem, that is, for complete distribution of class or label in the target column.

Table 4.1 Sample data

Day	Outlook	Temperature	Humidity	Wind	Play
1	Sunny	Mild	High	Strong	Yes
2	Rain	Cool	Normal	Strong	No
3	Sunny	Hot	High	Weak	Yes
4	Overcast	Cool	Normal	Strong	Yes
5	Rain	Mild	Normal	Weak	Yes
6	Sunny	Mild	High	Weak	Yes
7	Rain	Mild	High	Strong	No
8	Rain	Cool	High	Strong	Yes
9	Overcast	Hot	Normal	Weak	Yes
10	Overcast	Cool	Normal	Strong	Yes
11	Rain	Mild	Normal	Weak	Yes
12	Sunny	Mild	High	Weak	Yes
13	Sunny	Hot	Normal	Strong	No
14	Sunny	Mild	High	Weak	Yes
15	Rain	Mild	High	Weak	Yes
16	Overcast	Cool	Normal	Strong	Yes
17	Sunny	Mild	High	Weak	No
18	Sunny	Mild	High	Weak	Yes
19	Rain	Mild	Normal	Strong	No
20	Rain	Cool	High	Weak	No

The dataset has 14 positive instances (yes) and 6 negative (no) instances; therefore, $S=[14+,6-]$:

$$\text{Entropy}(S) = \text{Entropy}[14+,6-] = -\left(\frac{14}{20}\log_2\frac{14}{20} + \frac{6}{20}\log_2\frac{6}{20}\right) = 0.88$$

This concludes that the dataset is 88% impure or non-homogenous. Next, we measure the effectiveness of each attribute in classifying the training set. The measure we will use is called information gain or gain.

To make it clearer, let's use this formula of information gain and measure the gain of attribute wind from the dataset.

The attribute *wind* has the following values: weak or strong. Therefore, the distribution of class data in the *wind* attribute/feature would be:

- Wind(weak) $= W_W = [9+,2-]$, i.e., weak wind has 9 instances of yes and 2 instances of no.
- Wind(strong) $= W_S = [5+,4-]$, i.e., strong wind has 5 instances of yes and 4 instances of no.
- Entropy$(S) = 0.88$, i.e., entropy for complete sample space, which we calculated before.

Information gain for wind attribute can be expressed as the following:

$$\text{Gain}(S, \text{Wind}) = \text{Entropy}(S) - \left(\frac{11}{20}\text{Entropy}(W_W) + \frac{9}{20}\text{Entropy}(W_S)\right)$$

$$\text{Entropy}(W_W) = -\left(\frac{9}{11}\log_2\frac{9}{11} + \frac{2}{11}\log_2\frac{2}{11}\right) = 0.6840$$

$$\text{Entropy}(W_S) = -\left(\frac{5}{9}\log_2\frac{5}{9} + \frac{4}{9}\log_2\frac{4}{9}\right) = 0.9910$$

Now we substitute the values we found into the main formula and we get:

$$\text{Gain}(S, \text{Wind}) = 0.88 - \left(\frac{11}{20}0.6840 + \frac{9}{20}0.9910\right) = 0.88 - (0.3762 + 0.44595)$$

$$= 0.0578$$

So, the information gain by the *wind* feature is 0.057. Let's now similarly calculate the gain by the *outlook* feature.

For *outlook*, we have three values that are *sunny*, *rain*, and *overcast*. The distribution of class data in the *outlook* attribute/feature would be:

- Outlook(sunny) = O_S = [6+,2−]
- Outlook(overcast) = O_O = [4+,0−]
- Outlook(rain) = O_R = [4+,4−]

Information gain for *wind* attribute will be expressed as the following:

$$\text{Gain}(S, \text{Outlook}) = \text{Entropy}(S) - \left(\frac{8}{20}\text{Entropy}(O_S) + \frac{4}{20}\text{Entropy}(O_O) + \frac{8}{20}\text{Entropy}(O_R)\right)$$

$$\text{Entropy}(O_S) = -\left(\frac{6}{8}\log_2\frac{6}{8} + \frac{2}{8}\log_2\frac{2}{8}\right) = 0.8112$$

$$\text{Entropy}(O_o) = -\left(\frac{4}{4}\log_2\frac{4}{4} + \frac{0}{4}\log_2\frac{0}{4}\right) = 0$$

$$\text{Entropy}(O_R) = -\left(\frac{4}{8}\log_2\frac{4}{8} + \frac{4}{8}\log_2\frac{4}{8}\right) =$$

Now we substitute the values we found into the main formula, and we get:

$$\text{Gain}(S, \text{Outlook}) = 0.88 - \left(\frac{8}{20}0.8112 + \frac{4}{20}0 + \frac{8}{20}1\right)$$

$$= 0.88 - (0.3245 + 0 + 0.4) = 0.1555$$

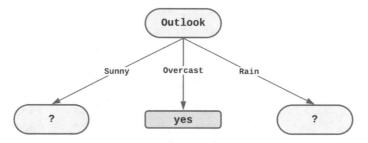

Fig. 4.4 Intermediate tree

Similarly, we will calculate information gain for the rest of the attributes/features, and we get the following results:

- Gain(outlook) = 0.1555
- Gain(wind) = 0.0573
- Gain(temperature) = 0.0019
- Gain(humidity) = 0.0018

The primary goal of information gain is to find the attribute/feature, which is most useful for classifying the data. The ID3 algorithm uses this attribute as its root to build the decision tree. Then it will repeatedly calculate *gain* to find the following node. Currently, the most useful attribute is *outlook* as it is giving more information than other attributes. So, *outlook* will be the root of our tree.

If we were to look at tree structure at this stage, it would look like the one shown in Fig. 4.4.

The *overcast* descendant has only *yes* instance and therefore becomes a leaf node with *yes*. For the other two nodes, the problem again arises which property/attribute should get tested. Now, these two nodes will be expanded by selecting the attribute/feature with the highest gain relative to the new subset of the tree.

Since the data points for the overcast are already classified, we won't use them for calculating other feature importance. We will utilize a subset of data points respective to features as seen in Table 4.2.

These are our rest of the data, which is yet to be classified.

Let's first find the attribute/feature that should get tested at the *sunny* descendant.

The subset of data of *sunny* descendants would include all data entry with the *sunny* as a feature as seen in Table 4.3.

Above are the data points with the *sunny* feature. The complete subset has eight entries with six instances of *yes* and two instances of *no*.

The sample space of this node would be:

$$S_{sunny} = S = [6+,2-]$$

Table 4.2 Subset of data points respective to features

Day	Outlook	Temperature	Humidity	Wind	Play
1	Sunny	Mild	High	Strong	Yes
2	Rain	Cool	Normal	Strong	No
3	Sunny	Hot	High	Weak	Yes
5	Rain	Mild	Normal	Weak	Yes
6	Sunny	Mild	High	Weak	Yes
7	Rain	Mild	High	Strong	No
8	Rain	Cool	High	Strong	Yes
11	Rain	Mild	Normal	Weak	Yes
12	Sunny	Mild	High	Weak	Yes
13	Sunny	Hot	Normal	Strong	No
14	Sunny	Mild	High	Weak	Yes
15	Rain	Mild	High	Weak	Yes
17	Sunny	Mild	High	Weak	No
18	Sunny	Mild	High	Weak	Yes
19	Rain	Mild	Normal	Strong	No
20	Rain	Cool	High	Weak	No

Table 4.3 Subset of data of sunny descendants

Day	Outlook	Temperature	Humidity	Wind	Play
1	Sunny	Mild	High	Strong	Yes
3	Sunny	Hot	High	Weak	Yes
6	Sunny	Mild	High	Weak	Yes
12	Sunny	Mild	High	Weak	Yes
13	Sunny	Hot	Normal	Strong	No
14	Sunny	Mild	High	Weak	Yes
17	Sunny	Mild	High	Weak	No
18	Sunny	Mild	High	Weak	Yes

The entropy for this subset would be given as:

$$\text{Entropy}(S) = \text{Entropy}[6+, 2-] = -\left(\frac{6}{8}\log_2\frac{6}{8} + \frac{2}{8}\log_2\frac{2}{8}\right) = 0.81$$

For this subset, we have a class entropy of 0.81%.

Let's start by calculating information gain for the *humidity* attribute for the taken subset of the tree.

Distribution for humidity in this case would be:

$\text{Humidity}_{\text{high}} = H_H = [6+, 1-]$
$\text{Humidity}_{\text{normal}} = H_N = [0+, 1-]$

Now we calculate information gain for this feature, by using the following equations:

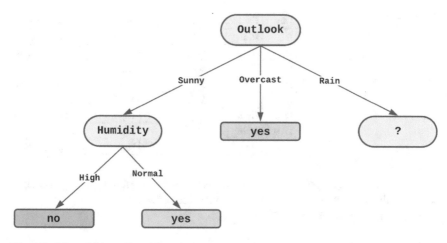

Fig. 4.5 After addition of humidity sub-tree

$$\text{Gain}(S, \text{Humidity}) = \text{Entropy}(S) - \left(\frac{7}{8}\text{Entropy}(H_H) + \frac{1}{8}\text{Entropy}(H_N)\right)$$

$$\text{Entropy}(H_H) = -\left(\frac{6}{7}\log_2\frac{6}{7} + \frac{1}{7}\log_2\frac{1}{7}\right) = (0.5915 + 0) = 0.5915$$

$$\text{Entropy}(H_N) = -\left(\frac{0}{1}\log_2\frac{0}{1} + \frac{1}{1}\log_2\frac{1}{1}\right) = (0 + 0) = 0$$

Now we substitute the values we found into the main formula, and we get:

$$\text{Gain}(S, \text{Humidity}) = \text{Entropy}(S) - \left(\frac{7}{8}*0.5915 + \frac{1}{8}*0\right) = 0.5175$$

Now measure the gain of *temperature* and *wind* by following the similar way as Gain(S, Humidity).

- Gain(S,Humidity) = 0.5175
- Gain(S,temperature) = 0.0725
- Gain(S,Wind) = 0.0725

The *humidity* attribute gives us the most information at this stage of iteration. The node after *outlook* at *sunny* descendant will be node *humidity*. In the *humidity* attribute, the *high* descendant has only *no* examples, and the *normal* descendant has only *yes* examples. So, both features *high* and *normal* can converge to classify target(leaf node) and cannot be further expanded.

At this stage, our constructed tree would look like the one seen in Fig. 4.5.

Since now, all the data points featuring *overcast*, *sunny*, and *humidity* types are covered in the following example. Now we decide the node for the *rain* descendant using the same operation we performed for the *humidity* subset. The subset of data of rain descendants would include all data entry with the *rain* feature as seen in Table 4.4.

Table 4.4 Subset of data of rain descendants

Day	Outlook	Temperature	Humidity	Wind	Play
D2	Rain	Cool	Normal	Strong	No
D5	Rain	Mild	Normal	Weak	Yes
D7	Rain	Mild	High	Strong	No
D8	Rain	Cool	High	Strong	Yes
D11	Rain	Mild	Normal	Weak	Yes
D15	Rain	Mild	High	Weak	Yes
D19	Rain	Mild	Normal	Strong	No
D20	Rain	Cool	High	Weak	No

Above are the data points with the *rain* feature. The complete subset has eight entries with four instances of *yes* and four instances of *no*.

The sample space of this node would be:

$$S_{rain} = S = [4+,4-]$$

The entropy for this subset would be given as:

$$\text{Entropy}(S) = \text{Entropy}[4+,4-] = -\left(\frac{4}{8}\log_2\frac{4}{8} + \frac{4}{8}\log_2\frac{4}{8}\right) = 1$$

Since all the data points associated with *outlook* (sunny and overcast) and *humidity* are classified. The only features left are *wind* and *temperature* in this case.

We will calculate information gain for these two features respective to the subset of the tree.

Now let's calculate information gain for attribute/feature *wind*.

Distribution for *wind* in the case of sub-tree would be:

$$\text{Wind}_{weak} = W_W = [3+,1-]$$
$$\text{Wind}_{strong} = W_S = [1+,3-]$$

Now we calculate information gain for this feature, by using following equation:

$$\text{Gain}(S, \text{Wind}) = \text{Entropy}(S) - \left(\frac{4}{8}\text{Entropy}(W_W) + \frac{4}{8}\text{Entropy}(W_S)\right)$$

$$\text{Entropy}(W_W) = -\left(\frac{3}{4}\log_2\frac{3}{4} + \frac{1}{4}\log_2\frac{1}{4}\right) = (0.3112 + 0.5) = 0.8112$$

$$\text{Entropy}(W_S) = -\left(\frac{1}{4}\log_2\frac{1}{4} + \frac{3}{4}\log_2\frac{3}{4}\right) = (0.5 + 0.3112) = 0.8112$$

We substitute the values we found into the main formula, and we get:

$$\text{Gain}(S, \text{Wind}) = \text{Entropy}(S) - \left(\frac{4}{8} * 0.8112 + \frac{4}{8} * 0.8112\right)$$

$$= 1 - (0.4056 + 0.4056) = 0.1888$$

Let's calculate information gain for attribute/feature *temperature*. Distribution for *temperature* in the case of sub-tree would be:

Temperature$_{hot}$ = T_H = [0+,0−]
Temperature$_{mild}$ = T_M = [3+,2−]
Temperature$_{cool}$ = T_C = [1+,2−]

Since in this subset, there are no instances of temperature *hot*, we will take 0 positive cases and 0 negative cases for this calculation.

We calculate information gain for this feature, by using the following equation:

$$\text{Gain}(S, \text{Temperature}) = \text{Entropy}(S)$$

$$- \left(\frac{0}{8} \text{Entropy}(T_H) + \frac{5}{8} \text{Entropy}(T_M) + \frac{3}{8} \text{Entropy}(T_C) \right)$$

$$\text{Entropy}(T_H) = -\left(\frac{0}{0} \log_2 \frac{0}{0} + \frac{0}{0} \log_2 \frac{0}{0} \right) = (0 + 0) = 0$$

$$\text{Entropy}(T_M) = -\left(\frac{3}{5} \log_2 \frac{3}{5} + \frac{2}{5} \log_2 \frac{2}{5} \right) = -(-0.4421 - 0.5287) = 0.9708$$

$$\text{Entropy}(T_C) = -\left(\frac{1}{3} \log_2 \frac{1}{3} + \frac{2}{3} \log_2 \frac{2}{3} \right) = -(-0.5253 - 0.3899) = 0.9182$$

We substitute the values we found into the main formula, and we get:

$$\text{Gain}(S, \text{Temperature}) = \text{Entropy}(S) - \left(\frac{0}{8} * 0 + \frac{5}{8} * 0.9708 + \frac{3}{8} * 0.9182 \right)$$

$$= 1 - (0 + 0.606 + 0.344) = 0.05$$

We got information gain for both the features:

- Gain(S,Wind) = 0.1888
- Gain(S,temperature) = 0.05

The *wind* attribute gives us the most information at this stage of iteration for construction. The node after *outlook* at *rain* descendant will be node *wind*. In the *wind* attribute, the weak descendant has only *yes* examples, and the strong descendant has only *no* examples. So, both features, weak and strong, can converge to classify the target(leaf node) and cannot be further expanded.

Since now, all the data points in the training set are classified using the following procedure, the final tree constructed would be as seen in Fig. 4.6.

This is how a decision tree is constructed using feature importance and measures of impurity technique.

With this detailed working of tree construction and traversal, let me summarize the algorithms for tree construction and traversal.

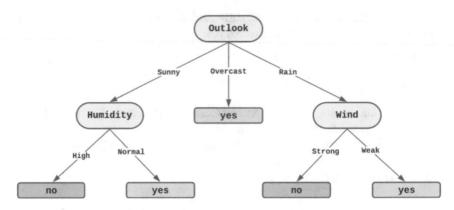

Fig. 4.6 Final tree

Tree Construction Algorithm

We express the complete process through the following algorithm:

- Select the feature that best classifies the dataset into the desired classes and assign that feature to the root node.
- Traverse down from the root node while making relevant decisions at each internal node such that each internal node best classifies the data.
- Route back to step 1 and repeat until you assign a class to the input data.

I will now explain you the tree traversal algorithm.

Tree Traversal Algorithm

You traverse the tree to visit a node in a tree or to print all values it contains. There are three ways to traverse a tree. These are called:

- In-order
- Pre-order
- Post-order

In the in-order traversal method, we visit the left sub-tree first, followed by the root and finally the right sub-tree. In the pre-order traversal, you visit the root first, followed by the left sub-tree and finally the right sub-tree. In the post-order traversal, we first visit the left sub-tree, followed by the right one and finally the root.

All three traversal methods are equivalent, and you may choose any of it with no further considerations. The effects would be the same in all three cases.

Implementation

The sklearn library provides the implementation of decision tree algorithms in two classes, one for regression *sklearn.tree.DecisionTreeRegressor* and the other for classification *sklearn.tree.DecisionTreeClassifier*.

```
class sklearn.tree.DecisionTreeClassifier(*,
        criterion='gini', splitter='best',
        max_depth=None, min_samples_split=2,
        min_samples_leaf=1,
        min_weight_fraction_leaf=0.0,
        max_features=None,    random_state=None,
        max_leaf_nodes=None, min_impurity_decrease=0.0,
        class_weight=None, ccp_alpha=0.0)
```

Both classes take many parameters; however, you will mostly use the default parameters set by the designers. Using default values for the parameters like *max_depth, min_samples_leaf*, etc. that control the size of the tree will cause a fully grown and unpruned tree. On large datasets, this will cause potentially very large trees. You will then need to set the values for these parameters to control the complexity and the size of the trees.

I will now present a trivial application to show how a decision tree algorithm is used.

Project (Regression)

I will use the famous Boston housing dataset provided in the UCI machine learning repository to estimate the price of your house in the Boston region.

Loading Dataset

As the dataset is available as a part of sklearn library, you load it into your project by simply calling the built-in function. After the dataset is loaded, create Pandas dataframe for further use.

```
<class 'pandas.core.frame.DataFrame'>
RangeIndex: 506 entries, 0 to 505
Data columns (total 14 columns):
 #    Column   Non-Null Count   Dtype
---   ------   --------------   -----
 0    CRIM     506 non-null     float64
 1    ZN       506 non-null     float64
 2    INDUS    506 non-null     float64
 3    CHAS     506 non-null     float64
 4    NOX      506 non-null     float64
 5    RM       506 non-null     float64
 6    AGE      506 non-null     float64
 7    DIS      506 non-null     float64
 8    RAD      506 non-null     float64
 9    TAX      506 non-null     float64
 10   PTRATIO  506 non-null     float64
 11   B        506 non-null     float64
 12   LSTAT    506 non-null     float64
 13   MEDV     506 non-null     float64
dtypes: float64(14)
memory usage: 55.5 KB
```

Fig. 4.7 Housing dataset information

```
house_dataset = load_boston()
house = pd.DataFrame(house_dataset.data,
                columns=house_dataset.feature_names)
house['MEDV']=house_dataset.target
```

Figure 4.7 shows the quick information summary of what it contains:

The dataset contains 14 columns. The first 13 will be used as features, and the last one MEDV is our target.

The description of each column is given below for your quick reference:

CRIM: Per capita crime rate by town
ZN: Proportion of residential land zoned for lots over 25,000 sq. ft
INDUS: Proportion of non-retail business acres per town
CHAS: Charles River dummy variable (= 1 if tract bounds river; 0 otherwise)
NOX: Nitric oxide concentration (parts per 10 million)
RM: Average number of rooms per dwelling
AGE: Proportion of owner-occupied units built prior to 1940
DIS: Weighted distances to five Boston employment centers
RAD: Index of accessibility to radial highways

TAX: Full-value property tax rate per 10,000
PTRATIO: Pupil-teacher ratio by town
B: $1000(Bk—0.63)^2$, where Bk is the proportion of (people of African American descent) by town
LSTAT: Percentage of lower status of the population
MEDV: Median value of owner-occupied homes in $1000s

Preparing Datasets

Next, you extract the features and the target.

```
X=house.drop('MEDV',axis=1)

y=house.MEDV
```

```
x_train, x_test, y_train, y_test = train_test_split(
                          X, y, test_size=0.2,
                          random_state=1234)
```

We have reserved 20% of the data for testing. At this point, our datasets are ready, and we will move to model building.

Model Building

You apply the decision tree regressor on the training dataset with the following code:

```
model=DecisionTreeRegressor(max_leaf_nodes=20,
                          random_state=123)
```

The *DecisionTreeRegressor* function takes several parameters; we have just used their default values. We train the model by calling its *fit* method.

```
model.fit(x_train,y_train)
```

The sklearn library also provides a function called *DecisionTreeClassifier* that you use if the target column has distinct values. In our case, the target has continuous values.

Evaluating Performance

After the model is trained, do the inference on the test dataset by calling the *predict* method and test the accuracy score:

```
y_pred=model.predict(x_test)
print('Accuracy on test set: {:.2f} %'.format(
     _model.score(x_test,y_test)*100))
```

It gave me around 87% accuracy on my test run. You may also check R2-score:

```
print("R-Squared = {}".format(r2_score(y_test, y_pred)))
```

which in my case was 86.79%. In regression, R-Squared is the most popular measure of goodness of fit. A value of R-Squared closer to 1 indicates better fitment.

Tree Visualization

You can get the tree visualization by calling the *plot_tree* method.

```
fig, axes = plt.subplots(nrows = 1,ncols = 1,
                             figsize = (4,4), dpi=300)
tree.plot_tree(model,
               feature_names = X.columns,
               class_names='MEDV',
               filled = True);
```

The tree visualization is shown in Fig. 4.8.

Feature Importance

Feature importance refers to techniques that assign a score to input features based on how useful they were while building a tree and at predicting a target variable. Feature importance is applied after the model is trained. You only "analyze" and observe which values have been more relevant in your model. Moreover, you will see that all *features_importances_* sums to 1, so the importance is seen as a percentage too. This also provides details on how the model selects features for particular tasks. You create feature importance plot using the following code snippet:

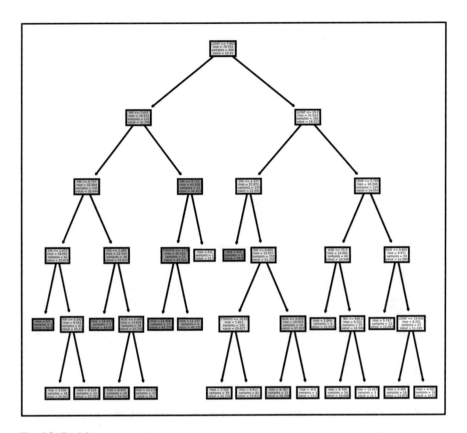

Fig. 4.8 Decision tree regressor

```
plt.figure(figsize=(10,8))
n_features = len(X.columns)#number of features
plt.barh(range(n_features), model.feature_importances_,
                align='center')
plt.yticks(np.arange(n_features), X.columns)
plt.xlabel("Feature importance")
plt.ylabel("Feature")
plt.ylim(-1, n_features)
```

The plot of features importance in my test run can be seen in Fig. 4.9.

As you see, the *LSTAT* feature has the highest importance in tree building, followed by *RM* as the second.

The full source of this project is available in the book's repository.

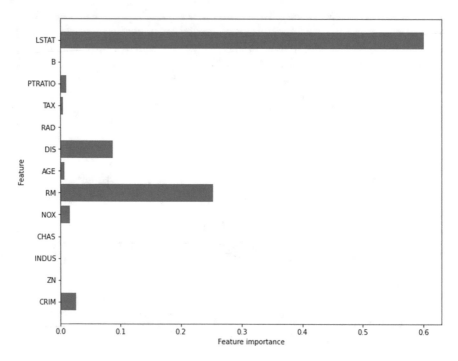

Fig. 4.9 Features importance

I will now show you the application of a decision tree classifier with the help of another trivial application.

Project (Classifier)

For this project, I have used the breast cancer dataset provided in the sklearn library. It is a small dataset (569 instances) and is thus ideally suited for visualizing decision trees. The dataset contains 30 features that determine the type of cancer (malignant or benign), which is going to be our target. So, this is a binary classification problem. I will show you the use of both gini and entropy approaches for building the classifier.

You load the dataset using the following code:

```
from sklearn import datasets
df = datasets.load_breast_cancer()
```

You can check the feature names by examining the value of *feature_names* attribute and labels by examining the value of *target_names*. The dataset is clean and well-balanced, so there was no need for any data preprocessing. After loading the dataset, I created the training/testing datasets in the ratio 70:30. Now comes the important part, defining the classifier.

You create the classifier using following code:

```
from sklearn.tree import DecisionTreeClassifier
classifier = DecisionTreeClassifier(
                    criterion='gini',
                    max_depth=3, random_state=0)
```

Note that I specified the criterion value to gini. Thus, the algorithm will use Gini index while building the tree. You build the decision tree by calling the *fit* method on the classifier instance.

```
classifier.fit(X_train, y_train)
```

Next, you will do the prediction on the test dataset and print the accuracy score as earlier. This was the score I got in my test run:

```
Accuracy score (gini): 0.9649
```

You visualize the created tree by calling the *plot_tree* method. Figure 4.10 shows the tree generated in my run.

Next, we will build the tree using an entropy approach. You create the classifier by specifying the criterion parameter to be entropy:

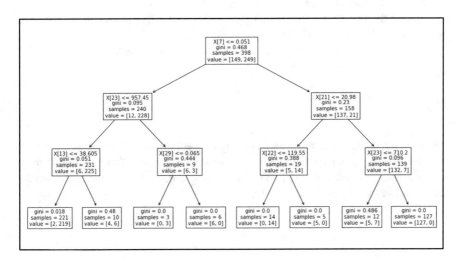

Fig. 4.10 Decision tree classifier (gini)

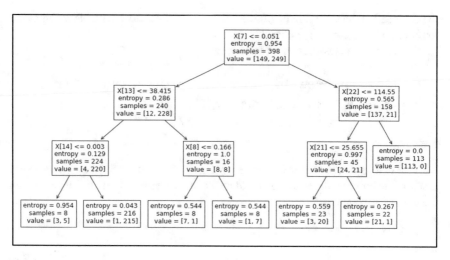

Fig. 4.11 Decision tree classifier (entropy)

```
classifier_entropy = DecisionTreeClassifier(
                        criterion='entropy',
                        max_depth=3, random_state=0)
```

I got the following accuracy score in my run:

```
Accuracy score (entropy) : 0.9474
```

You see, the accuracy achieved by both approaches is almost similar. Tree created by the entropy approach can be seen in Fig. 4.11.

Though the two approaches have produced trees of different structures, both will give out the similar results in inference of the unseen data.

The full source for this project is available in the book's repository.

Summary

Decision tree is an important algorithm in machine learning. We use it as a basis for several other more advanced algorithms. For this very reason, I covered this in so much detail. You learned how to build and traverse trees. We use the algorithm for both regression and classification tasks.

Chapter 5
Ensemble: Bagging and Boosting

Improving Decision Tree Performance by Ensemble Methods

In the previous chapter, you learned the decision tree algorithm. Building efficient trees is usually a complex, time-consuming process, especially so on high variance datasets. To improve the performance of decision trees, we use the statistical ensemble methods—bagging and boosting. Several such methods have been designed. In this chapter, you will learn the following methods.

- Random forest
- ExtraTrees
- Bagging
- AdaBoost
- Gradient boosting
- XGBoost
- CatBoost
- LightGBM

So, let us first understand the terms bagging and boosting.

What is Bagging and Boosting?

Bagging and boosting are the ensemble methods, in which the results of multiple models trained using the same algorithm are used as a final output. I will now describe both the methods.

Bagging

In case of bagging, we generate additional data for training from the same original dataset by using repetitions of some data points. We then train multiple models on

© The Author(s), under exclusive license to Springer Nature Switzerland AG 2023   97
P. Sarang, *Thinking Data Science*, The Springer Series in Applied Machine Learning,
https://doi.org/10.1007/978-3-031-02363-7_5

these datasets and combine them together to get us a last model. With boosting, we observe the observations of a classification model. If the observation is misclassified, we assign an additional weight to boost its importance. This is an iterative process, and, at the end, we get a stronger predictive model.

The data subsets that we require can be created using two different strategies:

- Bootstrap
- Out-of-bag

In bootstrap, the subset is created by replacing some data points with the randomly selected samples from the original. As a result, the subset would have the same number of data points as the original.

We train the same algorithm on different datasets created as above and take the average of all predictions, which will surely be more robust than a single classifier used on the full original dataset. This will effectively remove the variance in predictions observed over multiple runs using a single classifier.

In case of out-of-bag strategy, we create the subset by taking the difference between the original and the bootstrap datasets. Figure 5.1 shows how both bootstrap and out-of-bag datasets are created.

As you observe, each bootstrap dataset contains the same number of instances as the original set. In each bootstrap dataset, few data points are repeated. The corresponding out-of-bag datasets contain the varying number of instances. This depends on how many data points we repeated in the bootstrap dataset.

Boosting

In case of boosting, we start with a weak learner and strengthen it over multiple steps until it satisfies us with the classification results made by the last classifier. At each step, we look at the misclassification made by a hypothesis and increase its weight so that the next hypothesis will most likely classify it correctly. This is how we keep building stronger models, and thus the process is aptly called boosting.

With this brief introduction, I will now start describing the various algorithms, starting with random forest.

Random Forest

In a Nutshell

Instead of a one man decision, you decide based on decisions made by several. This is the principle used in random forest algorithms. Earlier, we saw the construction of a single decision tree on the entire dataset. This is like a one man decision for predictions on unseen data. Instead of that, let us make multiple trees by randomly

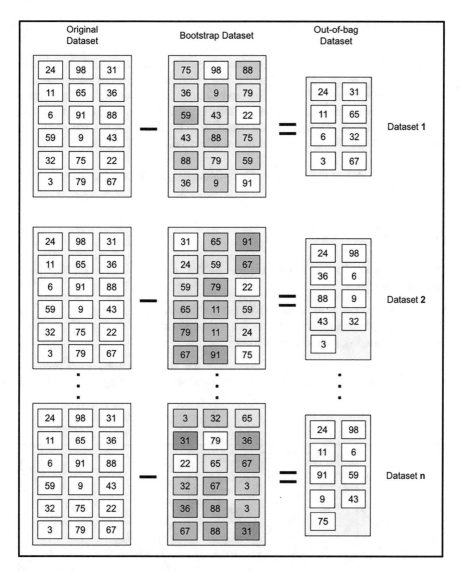

Fig. 5.1 Bootstrap and out-of-bag datasets

splitting the dataset into several parts. Now, let each decision tree make a prediction. The majority voting—the aggregation of predictions will decide the final prediction made by the full model.

What Is Random Forest?

Random forest is a supervised machine learning algorithm that can be used for solving both regression and classification problems. We base it on the concept of ensemble learning, which is like combining multiple classifiers to solve a complex problem. The random forest contains several decision trees built on various subsets of a dataset. Because of the majority voting, it improves the predictive accuracy of the model as compared to a single decision tree. Figure 5.2 illustrates the concept.

Now, let us look at the algorithm for building the random forest.

Random Forest Algorithm

Building a random forest is trivial. Here are the algorithmic steps for building a random forest:

1. Select K data points at random from the full dataset.
2. Build the decision tree on the selected K data points.
3. Repeat steps 1–2 for building N trees.

For predicting an unseen data point, use the votes made by each tree. For classification use majority voting and for regression use averaging.

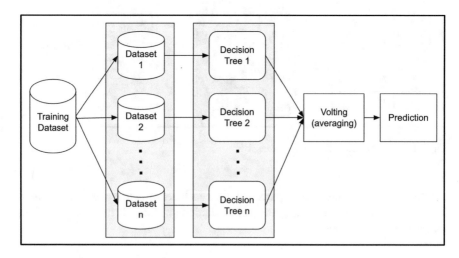

Fig. 5.2 Random forest working

Advantages

Here are a few advantages of this algorithm.

- Building trees is faster as each tree is constructed on a subset of the dataset with a subset of features.
- Algorithm is parallelizable, resulting in faster training.
- Gives better accuracy even on large datasets.

Applications

Data scientists have successfully used random forest algorithm in many sectors. In the banking sector, we use it for credit card fraud detection, customer segmentation, and predicting loan defaulters. In medicines, we use it in drug treatment classification, breast cancer diagnostics, cardiovascular disease prediction, and diabetes prediction. In e-commerce, we use it for product recommendation, sales predictions, price optimization, and search rankings. These are just a few examples. The random forest algorithm has found many use cases across several domains.

Implementation

The sklearn implements this algorithm in the class *sklearn.ensemble. RandomForesClassifier*. This is the way you invoke the algorithm:

```
model = RandomForestClassifier(n_estimators = 10,
               n_jobs = -1, criterion = "entropy",
               max_features = "auto",
               random_state = 1234)
model.fit(X_train, y_train)
```

The model takes several parameters; you may set the few essential parameters as shown in the above statement; the rest take their default values.

After the model is trained, you test its accuracy on the test data by calling its *score* method.

```
model.score(X_test, y_test)
```

Random Forest Project

To illustrate the use of the random forest algorithm, I have created a project. The project is exhaustive that also illustrates the use of random forest for *features* selection. It also shows you how to hyper-tune the algorithmic parameters. Rather than taking a huge, complex dataset, I have used the iris dataset to focus more on algorithm learning. The iris dataset is available as a part of sklearn library. Just for your quick recall, print the features and the target in the dataset.

These are the features:

```
['sepal length (cm)', 'sepal width (cm)', 'petal length (cm)', 'petal
width (cm)']
```

And these are the target names:

```
['setosa' 'versicolor' 'virginica']
```

As you see, there are three classes in the target variable. So, this is a multi-class classification task.

I did some univariate analysis on the dataset to check which features give a better separation of the classes. Figure 5.3 shows the class separations for the four features.

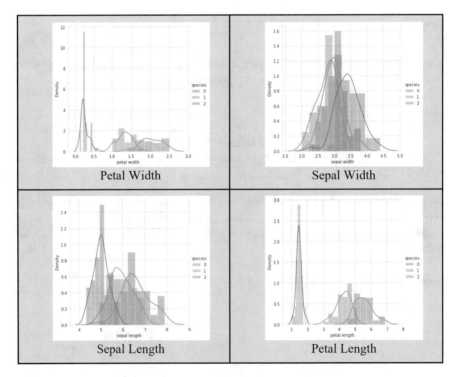

Fig. 5.3 Class distributions for various features

Observing the four charts, we see that the petal width probably provides a better separation of three classes.

After creating the training and testing datasets, we will apply the random forest algorithm using the following code snippet:

```
from sklearn.ensemble import RandomForestClassifier
clf=RandomForestClassifier(n_estimators=100)
clf.fit(X_train,y_train)
```

After the training, we will do the prediction on the test data and check the model's accuracy.

```
y_pred=clf.predict(X_test)

print("Accuracy:",metrics.accuracy_score(
                        y_test, y_pred))
```

This is the output in my run:

```
Accuracy: 0.9777777777777777
```

You can also print the classification report:

```
print("Classification re-port",
                metrics.classification_report(
                y_test, y_pred))
```

The output is shown in Fig. 5.4.

Random forest allows you to check the features importance by examining the value of an attribute as follows:

```
feature_imp = pd.Series(clf.feature_importances_,
                    index=iris.feature_names).sort_values(
                    ascending=False)
feature_imp
```

Classification report			precision	recall	f1-score	support
0	1.00	1.00	1.00	20		
1	1.00	0.91	0.95	11		
2	0.93	1.00	0.97	14		
accuracy			0.98	45		
macro avg	0.98	0.97	0.97	45		
weighted avg	0.98	0.98	0.98	45		

Fig. 5.4 Classification report

This is the output:

```
petal width (cm) 0.479486
petal length (cm) 0.418138
sepal length (cm) 0.079141
sepal width (cm) 0.023235
dtype: float64
```

You may also plot the features importance chart, as seen in Fig. 5.5.

As you observe, the petal width is the most important feature for class separation. This result tallies with our earlier manual inspection. The sepal width has the least impact on the classification. So, let us try dropping it—for dimensionality reduction.

```
# Removed feature "sepal width"
X1=data[['petal length', 'petal width','sepal length']]
y1=data['species']
X1_train, X1_test, y1_train, y1_test =
                       train_test_split(
                       X1, y1, test_size=0.3)
```

After dropping the column, retrain the model on the new dataset.

```
rf=RandomForestClassifier(n_estimators=100)
rf.fit(X1_train,y1_train)
```

This gives me the following accuracy score:

```
Accuracy: 0.9777777777777777
```

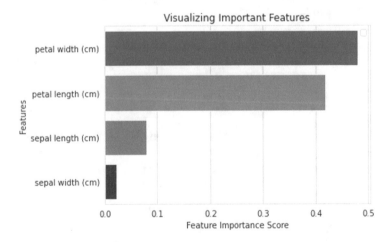

Fig. 5.5 Features importance in iris dataset

```
Classification report                     precision    recall  f1-score   support

                    0      1.00     1.00     1.00        18
                    1      0.93     1.00     0.96        13
                    2      1.00     0.93     0.96        14

          accuracy                          0.98        45
         macro avg      0.98     0.98     0.98        45
      weighted avg      0.98     0.98     0.98        45
```

Fig. 5.6 Classification report

The corresponding classification report is shown in Fig. 5.6.

We observe that the accuracy has improved after removing the least significant feature. Next, I will show you the technique of hyper-parameter tuning which we will use in the rest of the projects in this chapter. For parameter tuning, we use sklearn's *GridSearchCV* as shown in the code snippet below:

```python
from sklearn.model_selection import GridSearchCV

param_grid = {
    'bootstrap': [True],
    'max_depth': [80, 90, 100, 110],
    'max_features': ['sqrt', 'log2'],
    'min_samples_leaf': [3, 4, 5],
    'min_samples_split': [8, 10, 12],
    'n_estimators': [500, 700, 1000]
}
# Create a base model
clf = RandomForestClassifier()
# Instantiate the grid search model
grid_search = GridSearchCV(estimator = clf,
                param_grid = param_grid,
                cv = 3, n_jobs = -1, verbose = 2)
grid_search.fit(X_train,y_train)
```

We specify the various parameters that we wish to tune along with the desired test values in the *param_grid* variable. The *fit* method will take a long time to execute—this depends on the number of folds that you have set in the parameter list. After the execution, you can check the detected optimal values:

```python
grid_search.best_params_
```

Table 5.1 Model accuracies on different datasets

Dataset	Accuracy (%)
Full	97.78
Reduced	97.78
Full dataset with hyper-tuned parameters	96.00
Reduced dataset with hyper-tuned parameters	98.15

This is the output:

```
{'bootstrap': True,
 'max_depth': 90,
 'max_features': 'log2',
 'min_samples_leaf': 3,
 'min_samples_split': 12,
 'n_estimators': 500}
```

The *GridSearchCV* also returns the best model with these fine-tuned hyper-parameters. You can check its accuracy.

```
best_grid = grid_search.best_estimator_
grid_accuracy = evaluate(best_grid, X_test, y_test)
```

This is the output in my run:

```
Model Performance
Average Error: 0.0222 degrees.
Accuracy = 96.00%.
```

I also experimented with *GridSearchCV* on the reduced dataset. I observed the following accuracy score:

```
Model Performance
Average Error: 0.0222 degrees.
Accuracy = 96.67%.
```

Table 5.1 summarizes all the observations.

Note that these figures will change on each run.

The source for this project is available in the book's repository.

I will now discuss another bagging algorithm provided in sklearn library. It is called *ExtraTrees*.

ExtraTrees

The implementation of this algorithm is provided in *sklearn.ensemble. ExtraTreesRegressor* and *sklearn.ensemble.ExtraTreesClassifier* classes. The algorithm fits a number of randomized decision trees on various sub-samples and then uses averaging for improving predictive accuracy and controlling over-fitting.

To show its use, I created a small project that includes the previous random forest, *ExtraTrees*, and bagging algorithm as defined in sklearn. The bagging algorithm too has both regression and classification implementations and is discussed after this algorithm. By putting all three algorithms in a single project and implementing them on the same datasets gives us an opportunity to compare their performances. I will now describe the project.

Bagging Ensemble Project

I used the superconductivity dataset taken from the UCI repository for regression. It contains 21263 instances and has 80 features and 1 target. The target is the absolute temperature of the superconductor decided by the value of all features together. For classification, I used a molecular dataset again taken from the UCI repository. The dataset contains 21 columns (molecular attributes) of 1055 chemicals, and the label contains 2 classes with values, RB and NRB.

The algorithm provides a parameter called *n_estimators* to specify the number of trees in the forest. I experimented with a set of values for this parameter, so that you will understand how to get its optimal value. So, let us start with the regressor.

ExtraTreesRegressor

The following code snippet applies the regressor on our dataset for a set of *n_trees* values:

```
from sklearn.ensemble import ExtraTreesRegressor
ETreg = ExtraTreesRegressor(oob_score=True,
                            random_state=42,
                            warm_start=True,
                            n_jobs=-1,
                            bootstrap=True)
oob_list = list()
for n_trees in [15, 20, 30, 40, 50,
                100, 150, 200, 300]:
    ETreg.set_params(n_estimators=n_trees)
    ETreg.fit(X_train1, label_train1)
    oob_error = 1 - ETreg.oob_score_
    oob_list.append(pd.Series({'n_trees': n_trees,
                               'oob': oob_error}))

et_oob_df = pd.concat(oob_list,
                axis=1).T.set_in-dex('n_trees')
et_oob_df
```

The output will print the out-of-bag (oob) score for the various number of randomized trees on your terminal. I plotted these scores, which can be seen in Fig. 5.7.

Looking at the diagram, you will probably take 100 or 150 as the optimal value for the count of randomized trees.

To experiment the effect of other parameters, I tried three parameters listed in Table 5.2 and studied their *oob_errors*.

So, you may experiment with different values of these parameters or use *GridSearchCV* to get their optimal values.

You can do the prediction and print the classification report using the following code:

```
label_pred_reg_et=ETreg_300.predict(X_val1)
error_metrics(label_pred_reg_et,label_val1)
```

This is the output:

```
MSE: 82.74018764511439
RMSE: 9.096163347539136
Coefficient of determination: 0.920325118395126
```

I will now discuss the classifier implementation.

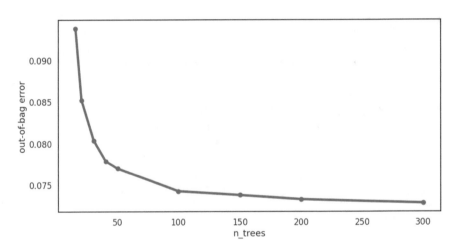

Fig. 5.7 Out-of-bag error vs number of trees

Table 5.2 The oob_error for various hyper-parameters

Hyper-parameter	oob_error
max_features='log2'	0.07450669925121
max_depth=3	0.29041541826443185
max_samples=0.7	0.07775480827253778

ExtraTreesClassifier

You apply the classifier on our classification dataset using the following code:

```
from sklearn.ensemble import ExtraTreesClassifier
ETcla = ExtraTreesClassifier(oob_score=True,
                             random_state=42,
                             warm_start=True,
                             n_jobs=-1,
                             bootstrap=True)
oob_list = list()
for n_trees in [15, 20, 30, 40, 50,
                100, 150, 200, 300]:
    ETcla.set_params(n_estimators=n_trees)
    ETcla.fit(X_train2, label_train2)
    oob_error = 1 - ETcla.oob_score_
    oob_list.append(pd.Series({'n_trees': n_trees,
                               'oob': oob_error}))

et_oob_df = pd.concat(oob_list,
                      axis=1).T.set_in-dex('n_trees')
et_oob_df
```

Figure 5.8 shows the result—*oob_error* for various values of randomized trees.

Like in the case of regressor, I studied the effect of two parameters. Table 5.3 shows the results.

Figure 5.9 shows the classification report in my run.

Overall, if you compare this with random forest, both algorithms have given us almost similar precision.

I will now discuss the third algorithm in this category and that is bagging.

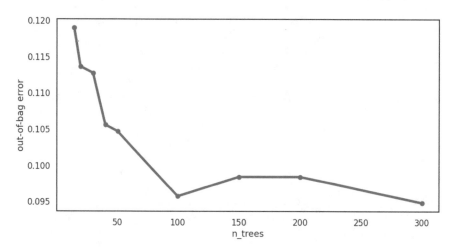

Fig. 5.8 The oob_error vs number of randomized trees

Table 5.3 The oob_error for different hyper-parameters

Hyper-parameter	oob_error
max_features='log2'	0.10286225402504467
max_depth=6	0.11270125223613592

```
               precision    recall  f1-score   support

           0       0.91      0.89      0.90       143
           1       0.89      0.91      0.90       137

    accuracy                           0.90       280
   macro avg       0.90      0.90      0.90       280
weighted avg       0.90      0.90      0.90       280
```

Fig. 5.9 Classification report

Bagging

The bagging algorithm allows you to set the base estimator of your choice. In the earlier two algorithms, the base estimator was a decision tree, which was non-changeable. I will show you the effect of different estimators on our datasets.

BaggingRegressor

I will use *SVR*, *KNeighbors*, and *Dummy* regressors as base estimators for testing. The sklearn library provides a dummy regressor, which is to be used only as a baseline for comparisons with other regressors. The following code snippet shows how to apply a bagging regressor with *SVR* as a base estimator:

```
from sklearn.ensemble import BaggingRegressor
from sklearn.svm import SVR
from sklearn.dummy import DummyRegressor
from sklearn.neighbors import KNeighborsRegressor

Bagreg_100 = BaggingRegressor(base_estimator=SVR(),
                              n_estimators=100,
                              oob_score=True,
                              random_state=42,
                              n_jobs=-1,
                              bootstrap=True)

Bagreg_100.fit(X_train1, label_train1)
oob_error100 = 1 - Bagreg_100.oob_score_
oob_error100
```

Table 5.4 The oob_score for different base estimators (regression)

Base estimator	oob_score
SVR	0.5260148888622115
DummyRegressor	1.0001160275212564
KNeighborsRegressor	0.11777280619044728

Table 5.4 shows the *oob_score*s for these three estimators.
The following output shows the error metrics produced by this regressor:

```
MSE: 126.49330021054058
RMSE: 11.246924033287527
Coefficient of determination: 0.8799453692899035
```

I will now discuss the classifier.

BaggingClassifier

I will use three estimators *SVC*, *Logistic regression*, and *KNeigbhors*. The following code snippet shows the use of *BaggingClassifier* with SVC as the estimator:

```
from sklearn.ensemble import BaggingClassifier
from sklearn.svm import SVC
from sklearn.linear_model import LogisticRegression
from sklearn.neighbors import KNeighborsClassifier

Bagcla_100 = BaggingClassifier(base_estimator=SVC(),
                               n_estimators=100,
                               oob_score=True,
                               random_state=42,
                               n_jobs=-1,
                               bootstrap=True)

Bagcla_100.fit(X_train2, label_train2)
oob_error100 = 1 - Bagcla_100.oob_score_
oob_error100
```

Table 5.5 gives the *oob_error*s with three different estimators.
Figure 5.10 is the classification report generated by the algorithm.
The for this project is available in the book's repository.
I will now discuss the boosting algorithms.

Table 5.5 The oob_error for different base estimators (classification)

Estimator	oob_error
SVC	0.1636851520572451
LogisticRegression	0.11627906976744184
KNeighborsClassifier	0.1520572450805009

```
               precision    recall   f1-score   support

           0        0.96      0.73       0.83        143
           1        0.78      0.97       0.86        137

    accuracy                            0.85        280
   macro avg        0.87      0.85       0.85        280
weighted avg        0.87      0.85       0.85        280
```

Fig. 5.10 Classification report

AdaBoost

This is an adaptive boosting classifier proposed by Yoav Freund and Robert Schapire in 1996. It iteratively ensembles multiple weak classifiers to build a strong classifier.

How Does It Work?

It is a meta-estimator that first fits a classifier on your dataset. On inference, you find that some data points are misclassified. The algorithm now focuses on these tough cases. After adjusting the weights, the classifier now makes another attempt; this time, the data points which were misclassified earlier would now be classified correctly. However, there may be other points which are misclassified. So, another iteration on the same dataset with newly adjusted weights would begin. This process continues until a pre-decided number of iterations or no further improvements in model building are observed.

AdaBoost, as you see, is a sequential ensemble process. The base estimator could be a decision tree or SVC (support vector classifier) or any other classifier that supports sample weighting.

I will now discuss its implementation as provided in sklearn library.

Implementation

The sklearn provides the AdaBoost implementation in *sklearn.ensemble. AdaBoostClassifier.*

```
class sklearn.ensemble.AdaBoostClassifier(
            base_estimator=None, *,
            n_estimators=50, learning_rate=1.0,
            algorithm='SAMME.R', random_state=None)
```

The class has only a few numbers of hyper-parameters. So, as a data scientist, it spared you of tweaking many parameters, resulting in considerable savings in development time.

You need to specify the base estimator, which by default is a decision tree classifier with maximum depth of 1, also known as decision stumps. You may specify the number of estimators at which boosting is ended, which by default is 50. In case of perfect fit, it automatically stopped the learning process early. The learning rate hyper-parameter decides the weight to apply to each iteration. A higher learning rate increases the contribution made by each classifier. There is a trade-off between the learning rate and the number of estimators. In the algorithm hyper-parameter, you may select between *SAMME* and *SAMME.R*. The latter converges faster.

Thus, you have a really limited number of hyper-parameters to tweak, which is an enormous advantage in using this algorithm. For tweaking, additionally you may use a *GridSearchCV* to get the optimal values for these parameters. I will now show you how to use the AdaBoost regressor.

AdaBoostRegressor

The following code snippet illustrates the implementation of AdaBoost regressor along with the use of grid search:

```
from sklearn.ensemble import AdaBoostRegressor
from sklearn.model_selection import GridSearchCV
Adareg = AdaBoostRegressor()

tuned_parameters = {'n_estimators': [50,100,200],
                'learning_rate':[0.001,0.01,0.1,0.5,0.9]}

Adareg_cv = GridSearchCV(Adareg, tuned_parameters,
                        cv=3,scoring='r2')
Adareg_cv.fit(X_train1, label_train1)
```

In the above code, we are evaluating three values for the number of estimators and five different learning rates. We use R-2 scoring for evaluation.

The *base_estimator* parameter by default is a decision tree. The number of estimators defines the number of stages, and the learning rate decides the weight applied at each iteration. There is a trade-off between the learning rate and the number of estimators. The higher learning rate would increase the contribution of each classifier. So, you may have to tweak these parameters to achieve your desired level of accuracy.

After the training completes, you can get the optimal values for the two hyper-parameters as determined by the *GridSearchCV*:

```
print(Adareg_cv.best_estimator_)
```

The output is:

```
AdaBoostRegressor(learning_rate=0.1, n_estimators=100)
```

Thus, the learning rate of 0.1 and the number of estimators equals 100 will give us the best performance on the model. This best performing model is already available to you as a returned object. Using this, you can do the prediction on the test data and evaluate the model's performance. This is the output on my test run:

```
MSE: 302.35435918758617
RMSE: 17.388339747876625
Coefficient of determination: 0.5555382513658986
```

The R-2 score (coefficient of determination) is not that good; this is probably because the model is not appropriate for our dataset. Let us now try the AdaBoost classifier.

AdaBoost Classifier

We will once again use the *GridSearchCV* for fine-tuning the hyper-parameters. The following code snippet shows the use of the classifier:

```
from sklearn.ensemble import AdaBoostClassifier

Adacla = AdaBoostClassifier(random_state=42)

tuned_parameters = {'n_estimators': [150,200,300],
                    'learning_rate':[0.4,0.5,0.7]}

Adacla_cv = GridSearchCV(Adacla, tuned_parameters,
                         cv=3,scoring='accuracy')
Adacla_cv.fit(X_train2, label_train2)
```

	precision	recall	f1-score	support
0	0.92	0.85	0.89	143
1	0.86	0.93	0.89	137
accuracy			0.89	280
macro avg	0.89	0.89	0.89	280
weighted avg	0.89	0.89	0.89	280

Fig. 5.11 Classification report

You can check the values of the hyper-parameters for this best performing model:

```
print(Adareg_cv.best_estimator_)
```

This is the output in my run:

```
AdaBoostClassifier(learning_rate=0.5, n_estimators=200,
random_state=42)
```

You can now do the predictions and print the classification report:

```
label_pred_cla_ada = Adacla_cv.predict(X_val2)
print(classification_report(label_val2,
                            label_pred_cla_ada))
```

The output is shown in Fig. 5.11.

The classification accuracy is probably within the acceptable range. We will try other algorithms to see if they give us better results.

Advantages/Disadvantages

These are some of the advantages offered by the algorithm:

- Easy to apply—only a few parameters to tweak.
- You may use any base classifier as long as it supports weighting.
- Not prone to over-fitting.
- Supports both classification and regression tasks.

A big disadvantage of AdaBoost is that it is sensitive to outliers and noisy data. The algorithm boosts their weights and tries to include them in the classification. I advise that you first clean up your dataset for outliers before applying the algorithm.

I will now discuss the next boosting algorithm, and that is *gradient boosting*.

Gradient Boosting

The word *gradient boosting* comprises two words gradient and boosting. You already know the meaning of boosting. The word gradient comes from gradient descent. If you have a numerical optimization problem where the aim is to minimize the loss function, then you use the *gradient boosting* algorithm. The gradient descent is a first-order iterative optimization algorithm for finding a local minimum of a differentiable function. You can apply *gradient boosting* to any of the well-known loss functions or your own created functions. The *gradient boosting* algorithm can apply to both regression and classification problems. When applied to decision trees for solving regression problems, it is sometimes called MART (multiple additive regression trees) or GBRT (gradient-boosted regression trees).

The working of *gradient boosting* can be explained with the help of a gradient descent diagram as shown in Fig. 5.12.

The algorithm starts with a single base learner or tree. At each iteration for some error or loss, it adds a new tree to reduce the error in the next stage. This process continues until there is no error or error is minimal. As you can see, the number of trees increases per iteration, reducing the error made by the model at each stage.

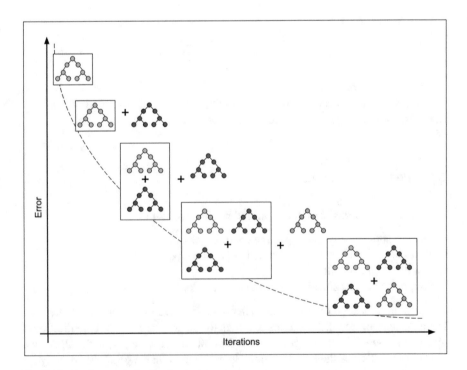

Fig. 5.12 Gradient optimization for decision trees

Loss Function

The loss function that you use in gradient descent depends on the problem you are trying to solve.

The sklearn library provides loss functions for both classification and regression problems. For classification, it supports two types of loss functions—*deviance* and *exponential*. For regression, it supports four loss functions—*squared_error*, *absolute_error*, *huber*, and *quantile*.

You may create your own loss function and set it as a parameter in a call to this algorithm.

Requirements for Gradient Boosting

Gradient boosting algorithm involves three elements:

- A loss function to be optimized
- Weak learner to make predictions
- An additive model to add weak learners for minimizing the loss function

As you have seen, the specific loss function depends on the kind of problem you are trying to solve. You may also use your own-defined function. The only requirement is that the function must be differentiable.

If your weak learner is a decision tree, which usually is the case, you must ensure that we construct the trees greedily. It helps in choosing the best split points based on purity scores, like Gini. Initially, you may use very short decision trees that only had a single split, called a decision stump. Decision stump is a unit depth tree which decides just one most significant cut on features. You may construct a weak learner by constraining the number of layers, nodes, splits, or leaf nodes. This is to ensure that the learners remain weak, but trees can be constructed in a greedy manner.

The last point, additive model, means that you should be able to add a single tree at each iteration, without affecting the existing trees at each stage. At each iteration, we calculate the loss or the error and rework the coefficients in a regression equation or if you are using this in neural networks, then the weights at the concerned layer in the network. At each iteration, we add a tree to the model that reduces the loss. We do this by parameterizing the tree. Simply change these parameters to move the tree in the right direction.

Next, I will discuss the sklearn implementation of this algorithm.

Implementation

The sklearn library provides the implementation of *gradient boosting* algorithm in these two classes—*sklearn.ensemble.GradientBoostingClassifier* and *sklearn. ensemble.GradientBoostingRegressor*. I will now show you how to implement these in our project.

GradientBoostingRegressor

You apply the gradient boosting regression algorithm to our dataset as shown in the code snippet below:

```
from sklearn.ensemble import GradientBoostingRegressor

GBR_reg = GradientBoostingRegressor(
                    learning_rate=0.4,max_depth=3,
                    n_estimators=300).fit(
                                X_train1, label_train1)
```

In the AdaBoost algorithm, I used *GridSearchCV* to get the optimal values of the parameters. You may do so in this and any of the following algorithms. For simplicity, I have used a few default values in the above code. After training, you do the predictions and print the error metrics using the following code:

```
label_pred_reg_gbr = GBR_reg.predict(X_val1)
error_metrics(label_pred_reg_gbr,label_val1)
```

This is the output in my run;

```
MSE: 104.8570274986513
RMSE: 10.239972045794428
Coefficient of determination: 0.9017534867290522
```

We observe that the R-2 score is far better than our previous algorithm—AdaBoost.

Now, let us try the classifier.

AdaBoostClassifier

You use the classifier using the following code snippet:

```
from sklearn.ensemble import GradientBoostingClassifier

GBC_cla = GradientBoostingClassifier(
                    learning_rate=0.5,max_depth=3,
                    n_estimators=200).fit(
                                X_train2, label_train2)
```

Again, I have used a few default values for the hyper-parameters. You do the predictions and print the classification report as follows:

```
label_pred_cla_gbc = GBC_cla.predict(X_val2)
print(classification_report(label_val2,
                            label_pred_cla_gbc))
```

The classification report generated in my run is shown in Fig. 5.13.

Compare this with the output of AdaBoost and you will notice that the results are far better than the earlier algorithm.

Pros and Cons

The gradient boosting algorithm can solve almost any objective function where we can write out a gradient. It means we can use it for ranking and Poisson regression, which are harder to achieve using random forest that you saw in the earlier chapter. The gradient boosting algorithm is a friendly approach to tackle multi-class problems that suffer from class imbalances.

The con side of the algorithm is that this is more sensitive to over-fitting, especially when the data is noisy. Second, the training may take a long time because trees are built sequentially. Lastly, for multi-class classification, as we decompose the problem into multiple binary-versus-all, it will cause a severe hit on the training time.

Next, I will discuss the *XGBoost* algorithm.

	precision	recall	f1-score	support
0	0.95	0.91	0.93	143
1	0.91	0.95	0.93	137
accuracy			0.93	280
macro avg	0.93	0.93	0.93	280
weighted avg	0.93	0.93	0.93	280

Fig. 5.13 Classification report

XGBoost

Released in 2014, *XGBoost* may be considered as further enhancements to the gradient boosting framework that you studied previously. As the name shows, it is an Extreme Gradient Boosting framework. It is an ideal blend of software and hardware optimization techniques. Mainly, it supports parallel distributed computing. It runs on a distributed environment like Hadoop Yarn, SGE (Sun Grid Engine), MPI, Apache Spark, Apache Flink, AWS EC2 cluster, and others. We can say that it is an optimized distributed gradient boosting library which is designed for efficiency, flexibility, and portability.

The *XGBoost* supports both L1 (Lasso) and L2 (Ridge) regularizations that help in preventing over-fitting. You decide on the desired regularization by passing the parameter (alpha and lambda) in sklearn's implementation. It supports parallel processing. The *nthread* hyper-parameter allows you to specify how many cores to use. If you do not specify its value, the algorithm will automatically detect the cores. The algorithm handles missing data values. When it encounters a missing value at a node, it tries both left and right splits to get the higher loss. It runs cross-validation in each iteration of the boosting process. This helps in achieving the optimal number of boosting iterations in a single run. Contrast this with the grid-searching algorithm used in gradient boosting machine (GBM). A GBM uses a greedy algorithm while splitting—it stops the split when a negative loss is encountered. On the other hand, *XGBoost* splits up to the specified maximum depth and then prunes the tree backward, removing splits beyond which there is no positive gain. *XGBoost* is an open-source library.

Here is the summary of its important features:

- Parallelized tree building
- Supports tree pruning (e.g., depth-first approach)
- Cache aware—data structures and code optimizations
- Implementation on various distributed systems
- Supports out-of-core computations
- Uses L1/L2 regularizations to avoid over-fitting
- Sparsity aware—handles missing data efficiently
- Employs weighted quantile sketch algorithm for optimal splits
- Supports cross-validation
- Implementations for Java, Scala, Julia, Perl, and other languages
- Linux, Windows, and macOS implementations
- Works for both regression and classification tasks

XGBoost is really fast when compared to other gradient boosting implementations prior to its introduction, as observed in many benchmark results.

A disadvantage would be that, like GBM, it does not perform well on sparse datasets. It is sensitive to outliers, as it forced every classifier to fix the errors in the preceding learners. Overall, it is not easily scalable, as each estimator bases its correctness on the preceding one.

Implementation

To use *XGBoost* in your applications, you will need to install it on your machine. You do so by calling *pip install* as follows:

```
!pip install xgboost
```

After the library is installed, creating both regressor and classifier models is trivial. I will first show you how to create a regression model.

XGBRegressor

The following code snippet calls the regressor on our dataset.

```
from xgboost import XGBRegressor

XGB_reg = XGBRegressor(
                    learning_rate=0.4,max_depth=3,
                    n_estimators=300).fit(
                            X_train1, label_train1)
```

The method takes several parameters. You may opt for their default values unless you want to really have control of the process and perhaps fine-tuning the performances. For the parameters that I have used in the above call, we could have used *GridSearchCV* as shown earlier to get their optimal values.

After the model training, we evaluate the performance using the following code:

```
label_pred_reg_xgb = XGB_reg.predict(X_val1)
error_metrics(label_pred_reg_xgb,label_val1)
```

This is the output on my test run:

```
MSE: 101.8290694199426
RMSE: 10.091039065425454
Coefficient of determination: 0.9047212957652594
```

Considering the fact that the model training is much quicker than the previous algorithms, the R-2 score is quite acceptable. Next, I will show you how to use the classifier.

```
              precision    recall  f1-score   support

         0       0.93      0.91      0.92       143
         1       0.91      0.93      0.92       137

  accuracy                          0.92       280
 macro avg       0.92      0.92      0.92       280
weighted avg     0.92      0.92      0.92       280
```

Fig. 5.14 Classification report

XGBClassifier

You use the classifier as shown in the code snippet here:

```
from xgboost import XGBClassifier

XGB_cla = XGBClassifier(
                 learning_rate=0.5,max_depth=3,
                 n_estimators=200).fit(
                          X_train2, label_train2)
```

You test the model's performance by printing the classification report.

```
label_pred_cla_xgb = XGB_cla.predict(X_val2)
print(classification_report(label_val2,
                          label_pred_cla_xgb))
```

The classification report generated in my run is shown in Fig. 5.14.

Considering the fact that our dataset size is just 1055 instances, the accuracy got is good.

Just to say a last word, rather than using *XGBoost* on small datasets, use it on really large datasets and you could see a noticeable difference in training times and also observe acceptable accuracy scores.

I will now discuss the next algorithm in our list, and that is *CatBoost*.

CatBoost

CatBoost is yet another open-source library that provides a gradient boosting framework. Originally developed at Yandex, it was open-sourced in July 2017 and is still in active development with Yandex and the community. The major enhancement, as compared to the other GB frameworks, is the support for categorical features. As a data scientist, this spares you from data preprocessing and converting those categorical columns to numeric.

Another important feature of this implementation is that it produces significant results even with default values of parameters. If you look up its documentation, you will see a very large number of parameters. Experimentation with these parameters would be a nightmare. Fortunately, the defaults assigned by the designers produce significant results in most situations.

Another major benefit of this framework is that it provides a fast prediction and thus is ideally suited for latency-critical applications. At Yandex, it is handling millions of users each month.

With their novel gradient boosting scheme, the designers were able to reduce the over-fitting. It also provides excellent support for GPUs for faster training and inference. Besides classification and regression tasks, it also supports ranking. It works on Linux, Windows, and macOS. It is available in Python and R. The models built using *CatBoost* can be used for predictions in C++, Java, C#, Rust, and other languages.

All said, it is practically being used in many applications developed by Yandex and companies like CERN and Cloudflare—to name a few it is used in search, recommendation systems, personal assistants, self-driving cars, and weather forecasting.

I will now show you how to use *CatBoost* in your projects.

Implementation

To install *CatBoost* on your machine, use *pip install* as follows:

```
!pip install catboost
```

First, I will discuss the regressor implementation followed by a classifier.

CatBoostRegressor

To apply the *CatBoost* on our dataset, use the following code:

```
from catboost import CatBoostRegressor

cat_model_reg = CatBoostRegressor(iterations=300,
                                  learning_rate=0.7,
                                  random_seed=42,
                                  depth=5)

cat_model_reg.fit(X_train1, label_train1,
                  cat_features=None,
                  eval_set=(X_val1, label_val1),
                  verbose=False)
```

Note that our dataset has no categorical features. If your dataset contains categorical features, list out the names of all your categorical variables as a value to the *cat_features* variable.

After the model training, do its performance evaluation using following two statements:

```
label_pred_reg_cat=cat_model_reg.predict(X_val1)
error_metrics(label_pred_reg_cat,label_val1)
```

Here is the output in my run:

```
MSE: 98.20372680432128
RMSE: 9.90977935194933
Coefficient of determination: 0.908991565534382
```

Next, I will demonstrate the classifier implementation.

CatBoostClassifier

You apply the classifier as follows:

```
from catboost import CatBoostClassifier

cat_model_class = CatBoostClassifier(iterations=300,
                                     learning_rate=0.7,
                                     random_seed=42,
                                     depth=3)

cat_model_class.fit(X_train2, label_train2,
                    cat_features=None,
                    eval_set=(X_val2, label_val2),
                    verbose=False)
```

As earlier, do the predictions and print the classification report:

```
label_pred_cla_cat = cat_model_class.predict(X_val2)
print(classification_report(label_val2,
                            label_pred_cla_cat))
```

Figure 5.15 shows the classification report produced in my run:

As you observe the performance is good and acceptable. Next, I will discuss the last algorithm in our list and that is *LightGBM*.

```
              precision    recall  f1-score   support

           0       0.96      0.90      0.93       143
           1       0.90      0.96      0.93       137

    accuracy                           0.93       280
   macro avg       0.93      0.93      0.93       280
weighted avg       0.93      0.93      0.93       280
```

Fig. 5.15 Classification report

LightGBM

Released in 2016, *LightGBM* is yet another open-source distributed gradient boosting framework originally developed by Microsoft. Considering the large datasets available these days, they developed it with the focus on performance and scalability. It carries almost all the features of *XGBoost*, the major difference being in the construction of trees. Conventionally, a tree is built row-by-row. *LightGBM* grows trees leaf-wise. A leaf with the probability of largest decrease in loss is selected during branching. Figure 5.16 gives a visual representation of the two schemes.

Most algorithms select the best split point based on the sorted features. Instead of this, the *LightGBM* uses a highly optimized histogram-based decision tree learning algorithm. This yields better efficiency and memory consumption.

The *LightGBM* algorithm also employs a novel technique called Gradient-Based One Side Sampling (GOSS). The GOSS works on the principle that the instances with larger gradients will contribute more to the information gain. Thus, it randomly drops the instances with small gradients while retaining the accuracy of the information. It also employs another novel feature called Exclusive Feature Bundling (EFB). This is a method for reducing the number of effective features. In a sparse dataset, many features are mostly exclusive. EFB bundles these features for a reduced dimensionality. The EFB, an exclusive feature bundle, is thus a bundle of these exclusive features.

Besides these differences, it provides all other advantages of *XGBoost*. It supports parallel training and the use of multiple loss functions. It supports sparse optimization. It has regularization and supports early stopping.

Briefly, its features may be summarized as follows:

- Low memory usage
- Distributed parallel training
- Supports GPU
- Provides better accuracy
- Ideally suitable for large datasets
- Faster training
- Supports categorical variables

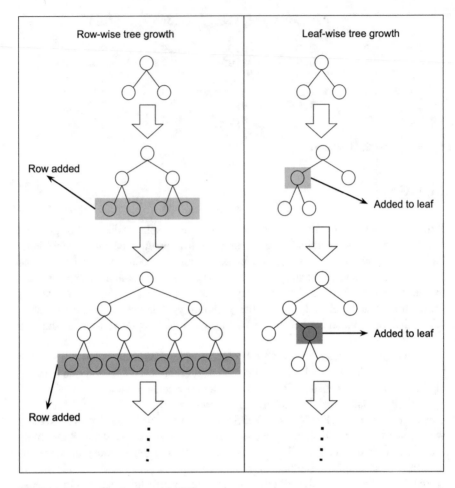

Fig. 5.16 Row and leaf-wise tree building

The disadvantages of *LightGBM* would be that generally due to its high sensitivity, it does not work well for small datasets, and it has too many parameters to play with.

I will now demonstrate its use.

Implementation

As in earlier cases, you install the library using *pip install*.

```
!pip install lightgbm
```

As earlier, I will demonstrate the use of both regressor and classifiers.

The LGBMRegressor

You apply the regressor to our dataset using the following code:

```
from lightgbm import LGBMRegressor

LGBM_reg = LGBMRegressor(iterations=100,

                         learning_rate=0.3,
                         depth=2)

LGBM_reg.fit(X_train1, label_train1,
             eval_set=(X_val1, label_val1),
             verbose=False)
```

You do the predictions and print the error metrics:

```
label_pred_reg_lgbm=LGBM_reg.predict(X_val1)
error_metrics(label_pred_reg_lgbm,label_val1)
```

This is the output in my run:

```
MSE: 89.05194394188753
RMSE: 9.436733753894275
Coefficient of determination: 0.9170497217299074
```

As you observe, it has given us good results. Next, I will demonstrate the use of a classifier.

The LGBMClassifier

You use the classifier using the following code:

```
from lightgbm import LGBMClassifier

LGBM_cla = LGBMClassifier(iterations=300,
                          learning_rate=0.4,
                          depth=3)

LGBM_cla.fit(X_train2, label_train2,
             eval_set=(X_val2, label_val2),
             verbose=False)
```

You do the model evaluation and print the classification report as follows:

```
label_pred_cla_lgbm = LGBM_cla.predict(X_val2)
print(classification_report(label_val2,
                            label_pred_cla_lgbm))
```

The classification report generated in my run is shown in Fig. 5.17.

As I have completed the discussion of the different boosting algorithms, I will now summarize our observations.

Performance Summary

In my summary report, I have also included random forest. Like other algorithms, this was tested for both regression and classification. I used the same datasets for all the tests. Figure 5.18 shows the summary for regression.

For the dataset that we have used, random forest has given better accuracy as compared to others. This is mainly because of the dataset size, which is too small for other algorithms. The later algorithms provide additional advantages for really large datasets. Figure 5.19 shows the regression plot generated by random forest.

Let us now look at the classification comparisons. Figure 5.20 gives a summary of the classification report for various algorithms.

As you can see, *CatBoost* is a winner in this case. Figure 5.21 presents the confusion matrix generated by *CatBoost*.

The full source code for this project is available in the book's repository.

	precision	recall	f1-score	support
0	0.93	0.92	0.93	143
1	0.92	0.93	0.92	137
accuracy			0.93	280
macro avg	0.92	0.93	0.92	280
weighted avg	0.93	0.93	0.93	280

Fig. 5.17 Classification report

	Random Forest	AdaBoost	Gradient Boosting	XGBoost	CatBoost	LightGBM
MSE	80.507403	308.295598	104.854045	101.829069	98.203727	89.051944
RMSE	8.972592	17.558348	10.239826	10.091039	9.909779	9.436734
R2 score	0.923425	0.504163	0.901580	0.904721	0.908992	0.917050

Fig. 5.18 Summary report for regression

Fig. 5.19 Regression plot
generated by random forest

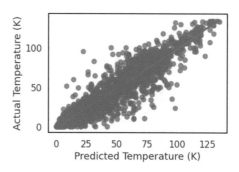

	Random Forest	AdaBoost	Gradient Boosting	XGBoost	CatBoost	LightGBM
precision	0.925033	0.891884	0.929425	0.918087	0.934026	0.925033
recall	0.925000	0.889286	0.928571	0.917857	0.932143	0.925000
fscore	0.925005	0.889208	0.928571	0.917867	0.932119	0.925005
accuracy	0.925000	0.889286	0.928571	0.917857	0.932143	0.925000
auc	0.925042	0.890077	0.928998	0.918049	0.932801	0.925042

Fig. 5.20 Summary report for various algorithms

Fig. 5.21 Confusion matrix
for CatBoost

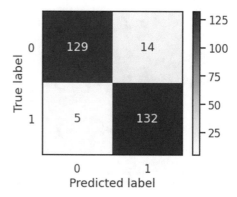

Summary

In this chapter, you studied the statistical ensemble techniques for boosting the performance of a decision tree classifier. You studied both bagging sand boosting techniques. Several bagging and boosting techniques were developed over a period, each adding up the improvements to the earlier one. It does not mean that you should use only the latest one. As you have seen, the advanced techniques also require larger datasets. So, as a data scientist, you may begin with a simple decision tree or random forest, and if it cannot give you your desired results, then move on to the higher models.

Chapter 6
K-Nearest Neighbors

A Supervised Learning Algorithm for Classification and May Be Regression

In this chapter, you will learn the use of another important machine learning algorithm that is the *k*-nearest neighbors.

In a Nutshell

Let me explain it with a trivial example. Consider you meet a new person. With brief hesitation, you say he is Chinese. How do you conclude so easily that the person is of Chinese origin? With your previous knowledge of the human anatomy of persons of different origins, you make this decision. Basically, you try to match the distinct features of his body with all previously known person's features. For example, you may look at his eyes, his body-built, height, complexion, and so on. If many of these features come close to the category of people known to you previously as Chinese, you conclude that this new person is most likely to be Chinese.

We use a similar logic in *k*-nearest neighbors. To compute the similarity of features, we must represent each of these features in numeric values. The features like weight and height are numeric. But what about eyes? We may convert these into numeric values by considering the height/width of an eye rather than saying the eye is big or small. Thus, if you can convert those "important" features into numeric data, you can use the KNN algorithm to train the machine. We can then use such a trained ML model for classification and clustering. Though this discussion pertains only to classification, we can use the algorithm even for solving regression problems. Note that this is a supervised learning algorithm, so you will need a labeled dataset.

With this overview of the algorithm, let us look at its details.

© The Author(s), under exclusive license to Springer Nature Switzerland AG 2023
P. Sarang, *Thinking Data Science*, The Springer Series in Applied Machine Learning, https://doi.org/10.1007/978-3-031-02363-7_6

K-Nearest Neighbors

KNN is one of the easiest algorithms to implement. The algorithm stores all data and classifies a new data point based on its similarity to other data points. Though we mostly use it for classification, we can also use it for regression analysis. It is a non-parametric algorithm in the sense that it makes no assumptions about the underlying data.

The diagram shows the result of KNN classification. Given a set of data points (news articles), it classifies the documents into three categories, sports, politics, and finance. This is depicted in Fig. 6.1.

Note that we must convert the text documents into vectors before they can be input into the algorithm. After we classify the training dataset into different categories, the category of an unseen data point can be determined by finding its closeness to the neighboring data points. For example, the data point in question (black rectangle) has three close blue points and two red points, followed by one green point in that order. As the number of blue points is more than compared to the other two, the data point would be classified as belonging to the blue group, that is, the finance category.

I will now show you the algorithm.

KNN Algorithm

- Step-1: Load the data.
- Step-2: Initialize *K* to your chosen number of neighbors, five as an example.

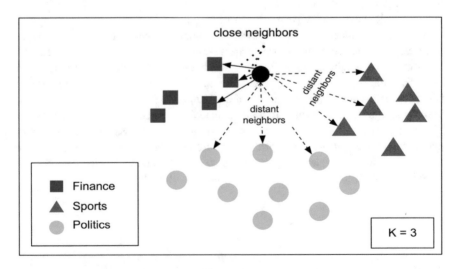

Fig. 6.1 Close and distant neighbors—the black point classifies as finance

- Step-3: For each data point in the dataset:

  - Calculate its distance from every other point in the entire dataset.
  - Store the distance and the index of each such element (data point) into an ordered collection.

- Step-4: Sort the collection into the ascending order of distances.
- Step-5: Pick the first K entries from the sorted collection.
- Step-6: Get the labels of the selected K entries.
- Step-7: For classification, use the statistical mode value of the selected K labels as a prediction. For regression, use the statistical mean.

I will now show you the working of this algorithm with a few illustrations.

KNN Working

Consider our earlier example of identifying the person's origin by examining his four features, eye diameter, height, weight, and complexion. Table 6.1 shows the sample labeled data.

Now, consider that you want to identify the person with the following data point:

| 0.9 | 6.2 | 198.4 | 0.8 | Unknown |

You will now compute the Euclidean distance of this unknown person with every other person in the dataset. Figure 6.2 shows the computation for the first person.

Figure 6.3 shows the computation of the second person.

Now, you have unsorted collection of Euclidean distances of all data points as shown in Table 6.2.

Sort this collection in the descending order. The sorted output is shown in Table 6.3.

Now, take the top 5 entries. Why 5? Because we want to consider the top 5 predictions. In other words, we want to set $K = 5$. I will explain to you what happens when you take a different value for K. What is the majority voting (mode in statistical terms) among these top 5 entries? It is E_3. What is the class for E_3? It is

Table 6.1 Sample dataset for KNN

Eye diameter (Inches)	Height (feet)	Weight (Pound)	Complexion (fairness)	Origin
0.8	6.2	176.4	1	American
0.6	5	143.3	0.9	Japanese
1	5.6	165.3	0.5	Indian
0.5	4.10	110.2	0.8	Chinese
0.9	6	156.4	1	British
1	6.5	220.5	0	African
0.9	6.2	198.4	0.8	Australian

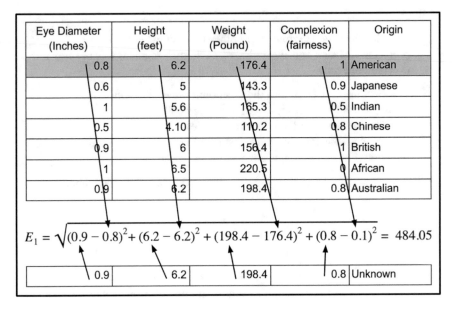

Fig. 6.2 Euclidean distance computations for first data point

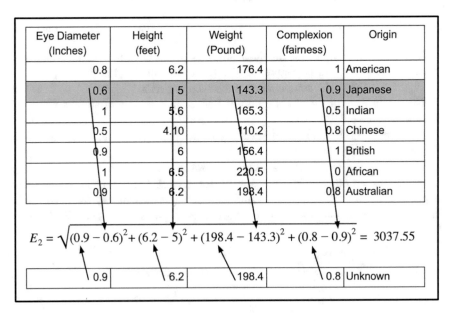

Fig. 6.3 Euclidean distance computations for second data point

Table 6.2 Euclidean distances for all data points

E_1	484.05
E_2	3037.55
E_3	1096.07
E_4	7783.81
E_5	1764.53
...	
E_n	1796.18

Table 6.3 Sorted Euclidean distances

E_4	7783.81
E_2	3037.55
E_n	1796.18
E_5	1764.53
E_3	1096.07
E_1	484.05

Indian. So, you can conclude that the unknown person is of Indian origin. This is how the KNN algorithm works.

You may also use KNN for regression analysis. Here, you will use the mean value of the top K entries as your predicted output.

I will now explain to you what happens when you select a different value for K.

Effect of K

Consider the following visual depiction, as shown in Fig. 6.4, of an unknown data point in a cluster of known data points.

As seen in Fig. 6.4, if you take the value of $K = 5$, there are two red (class B) objects and three blue (class A) objects in the vicinity of the unknown object. So, you will classify the unknown object as of type class A, considering the majority voting or the *mode* value in statistical terms. However, if you take K equals 10, you have 6 red objects (class B) and 4 blue (class A) objects. Here, you will classify the unknown object as of type class B (red). Thus, you can see how the classification can vary depending on the selection of K.

I will now discuss the advantages and disadvantages of the KNN algorithm.

Advantages

Here are some of the advantages:

- Simple to implement
- Robust to the noisy training data
- Can be more effective for large datasets

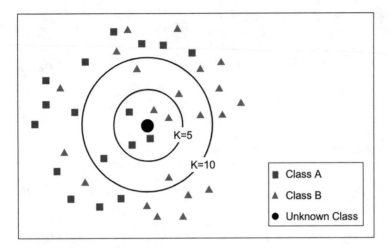

Fig. 6.4 Effect of *K* in class determination

Disadvantages of KNN

Here are two major disadvantages of KNN:

- An appropriate selection of *K* value can be tricky.
- Computation cost is high as you need to calculate the distance between the unknown point and all other points in the entire dataset.

Let us now look at the implementation of this algorithm as provided in sklearn library.

Implementation

The sklearn library provides the implementation of KNN algorithm in *sklearn. neighbors.KNeighborsClassifier*. While classifying, it implements the *k*-nearest neighbors vote. The class is declared as follows:

```
class sklearn.neighbors.KNeighborsClassifier(
          n_neighbors=5, *, weights='uniform',
          algorithm='auto', leaf_size=30, p=2,
          metric='minkowski',
          metric_params=None, n_jobs=None)
```

When you create a class instance, you will need to set values for these parameters (or accept defaults). The *n_neighbors* parameter decides the number of neighbors to be used during voting. The *weights* parameter whether all points in each neighborhood are weighted equally or closer neighbors will have a greater influence than those who are further away. You may also provide a user-defined function as an argument here to get a control on who is your neighbor, The algorithm parameter gives you a choice of algorithms to use, which can be *BallTree*, *KDTree*, or brute-force search. The default value for this parameter is *auto*, which attempts to find the most appropriate algorithm for your dataset. So, rarely, you will need to set this on your own. For other parameters, you can simply accept their default values.

Ultimately, as a data scientist, you need to set only the value of *n_neighbors*. The question is how to decide on the optimal value of this parameter. To understand this and how KNN is applied, I will present a real-life example.

Project

Classifying medical diagnostic data of thousands of people into diabetic and non-diabetic categories can help medical practitioners in early detection of diabetes in their patients. Fortunately, the National Institute of Diabetes and Digestive and Kidney Diseases has made such a database available for machine learning. We will apply our KNN algorithm on this dataset to classify the data into two categories. The dataset that I am going to use is a subset of the larger database and contains data points pertaining to females of Pima Indian heritage older than 21 years. The dataset contains many vital parameters, such as number of previous pregnancies, BMI, insulin level, age, and so on. So, this is a multifeature dataset with two output categories.

Loading Dataset

I have kept the dataset in the book's repository for easy access to this project. Download the dataset using the *wget* command:

```
!wget 'https://raw.githubusercontent.com/prof-
sarang/Springer/main/diabetes.csv'
```

Load the data into a Pandas dataframe and get its info using the following code:

```
df = pd.read_csv('/content/diabetes.csv')
df.info()
```

The information of the dataset can be seen in Fig. 6.5.

The *Outcome* is our target column, and the rest of the columns will be our features. Also, notice that we have just 768 data points. Having 8 predictors, a larger number of data points would have helped us to get better accuracy on KNN algorithms.

```
<class 'pandas.core.frame.DataFrame'>
RangeIndex: 768 entries, 0 to 767
Data columns (total 9 columns):
 #    Column                    Non-Null Count   Dtype
---   ------                    --------------   -----
 0    Pregnancies               768 non-null     int64
 1    Glucose                   768 non-null     int64
 2    BloodPressure             768 non-null     int64
 3    SkinThickness             768 non-null     int64
 4    Insulin                   768 non-null     int64
 5    BMI                       768 non-null     float64
 6    DiabetesPedigreeFunction  768 non-null     float64
 7    Age                       768 non-null     int64
 8    Outcome                   768 non-null     int64
dtypes: float64(2), int64(7)
memory usage: 54.1 KB
```

Fig. 6.5 Dataset information

We now extract features and targets into *X* and *y* dataframe.

```
X = df.drop('Outcome',axis=1).values
y = df['Outcome'].values
```

We split the entire dataset into training and testing using the *train_test_split* method of sklearn.

```
X_train,X_test,y_train,y_test = train_test_split(
                                X,y,test_size=0.2,
                                random_state=42,
                                stratify=y)
```

We reserve 20% of the data for testing.

At this point, we are not ready to train our KNN algorithm. First, we will determine the optimal value for *K*.

Determining **K** *Optimal*

To determine the optimal value of *K*, we use a simple trick. We apply the classifier on the same dataset with different values of *K*. We use the default values for all other parameters to the classifier. We do this using the following code fragment:

```
#Setup arrays to store training and test accuracies
neighbors = np.arange(1,15)
train_accuracy =np.empty(len(neighbors))
test_accuracy = np.empty(len(neighbors))

for i,k in enumerate(neighbors):
    #Setup a knn classifier with k neighbors
    knn = KNeighborsClassifier(n_neighbors=k)

    #Fit the model
    knn.fit(X_train, y_train)

    #Compute accuracy on the training set
    train_accuracy[i] = knn.score(X_train, y_train)

    #Compute accuracy on the test set
    test_accuracy[i] = knn.score(X_test, y_test)
```

We plot the accuracies for our visualization. Such a plot is shown in Fig. 6.6.

You observe that the accuracy on the training dataset starts flattening with $K = 4$, though on the test data, a few kinks continue. We will take $K = 4$ for our final training.

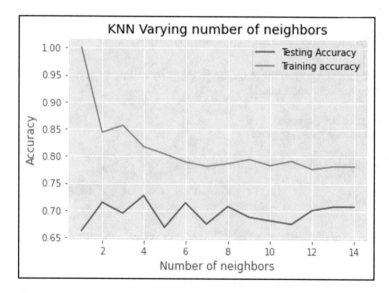

Fig. 6.6 Accuracy vs. number of neighbors for KNN

Model Training

We create the classifier with $K = 4$.

```
#Setup a knn classifier with k neighbors
knn = KNeighborsClassifier(n_neighbors=4)
```

We train the model on the entire dataset.

```
#Fit the model
knn.fit(X_train,y_train)
```

We check the model's accuracy score:

```
knn.score(X_test,y_test)
```

The output is:

```
0.7272727272727273
```

The accuracy is not that good, mainly due to the limited number of data points that we have in this dataset.

Model Testing

We evaluate the model's performance on the test data by calling its *predict* method and then plotting the confusion matrix:

```
y_pred = knn.predict(X_test)
confusion_matrix(y_test,y_pred)
```

Here is the output:

```
array([[90, 10],
 [32, 22]])
```

You may also print the entire classification report to examine the different scores like precision, recall, f1-score, and support.

```
print(classification_report(y_test,y_pred))
```

The classification report is shown in Fig. 6.7.

The full source of this project is available in the book's repository. I will now describe a few application areas where you can use this algorithm.

	precision	recall	f1-score	support
0	0.74	0.90	0.81	100
1	0.69	0.41	0.51	54
accuracy			0.73	154
macro avg	0.71	0.65	0.66	154
weighted avg	0.72	0.73	0.71	154

Fig. 6.7 Classification report for KNN

When to Use?

There are several domains and use-cases where the KNN algorithm has been successfully applied. We use it for estimating credit score for a new applicant based on the information on the past creditors. It is used to determine the outliers in credit card fraud detection systems. We use it to identify semantically similar documents where each document is considered a vector for machine learning. We can also use it for text categorization, which is a classification problem. Though KNN is not suitable for high-dimensional data, we can use it to create a baseline for the recommendation systems.

Another example for its use as a classifier: consider the case of identifying fields in an MS Excel sheet. Different persons in an organization submit a similar Excel sheet where the field names differ to some extent. KNN can automatically detect the similarity of data in such fields and group them together. This is a very useful technique, as most of the time the users of Excel sheet do not adhere to a specific template. This can be very useful when you do data transformation.

To give you an example of a regression analysis, we can use KNN for estimating a person's weight given his height and age by finding the k-nearest neighbors and then applying the mean value of age to this unknown object.

Summary

The KNN algorithm uses the similarity measures to decide upon the k-nearest neighbors. After we store the similarity measures for each data point, we can classify an unseen data point as of a certain type based on its similarity with the other data points in the entire dataset. You studied the sklearn implementation of this algorithm. The only parameter that you need to worry about while using the algorithm is the value for k-neighbors. You studied a technique for determining an optimal value for this k-neighbor.

In the next chapter, you will study naive Bayes algorithm.

Chapter 7
Naive Bayes

A Supervised Learning Algorithm for Classification

In this chapter, you will learn the use of another important machine learning algorithm that is based on naive Bayes theorem.

In a Nutshell

When you want to make quick predictions on a high-dimensional dataset, you use naive Bayes. This is one of the most efficient algorithms for classification and probably the simplest. When you have several thousand data points and many features in your dataset, it trains quickly to help you get predictions in real time. It thus helps in building the fast machine learning models to make quick predictions. It is also easy to build.

Naive Bayes theorem makes a "naive" assumption that the various features are conditionally independent of each other, which may not be always true. Despite this assumption, we observe that the algorithm outperforms even highly sophisticated classification methods. It is a probabilistic classifier. It considers every feature to contribute independently to the probability of occurrence of the target value, irrespective of the correlation to other features. With this naive (simple) assumption, it predicts using the maximum likelihood. The algorithm is more suitable for supervised learning. The Bayesian methods, which are derived from the principles of Bayesian inference, are good for supervised learning. Bayesian inference is a statistical inference method in which we update the probability for a hypothesis as more information becomes available.

© The Author(s), under exclusive license to Springer Nature Switzerland AG 2023
P. Sarang, *Thinking Data Science*, The Springer Series in Applied Machine Learning,
https://doi.org/10.1007/978-3-031-02363-7_7

When to Use?

A few applications of this algorithm are spam filtering, sentiment analysis, and classifying journal articles. To cite a concrete example, consider a model that helps you in deciding when to go out for a play. Going out depends on factors such as outside temperature, rain, and humidity. All these factors independently contribute to the probability of playing out, even though in nature they depend on each other. Thus, the algorithm is called "naive" due to its simplicity assumption.

To understand the implementation of this algorithm, let us have a quick look at naive Bayes theorem.

Naive Bayes Theorem

Bayes' theorem, also known as Bayes' rule or law, is used for determining the probability of a hypothesis with prior knowledge. The formula is shown in Fig. 7.1.
   where:

- $P(c|x)$ is the posterior probability of class c given predictor x
- $P(c)$ is the prior probability of class c
- $P(x|c)$ is the probability (likelihood) of a predictor given class c
- $P(x)$ is the prior probability of predictor x

The final posterior probability of class c is computed taking into account all the features as follows:

$$P(c|X) = P(x_1|c) * P(x_2|c) * \ldots * P(x_n|c) \times P(c)$$

where $x_1, x_2, \ldots, x_n$ are the predictors and there are n predictors.
   I will now illustrate how these probabilities are computed for a real dataset.

Fig. 7.1 Bayes' theorem

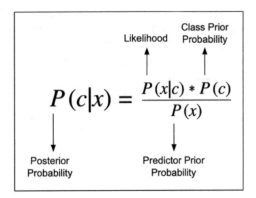

Applying the Theorem

To understand how the theorem is applied, I will present a trivial case study. Consider a bank that gives loans to eligible candidates. The loan eligibility depends on several factors (features). We use the naive Bayesian equation to calculate the posterior probability for each class. The class with the highest posterior probability will be the prediction made by the model.

To keep it simple, I will show you the probability of computation on just one field—the customer's education. Table 7.1 shows sample data:

Figure 7.1 shows the loan status—if approved or not and the corresponding education status of the applicant.

We use this formula for calculating the posterior probability for education.

Formula

$P(c|x) = p(x|c) * p(c)/p(x)$

For our example we consider $c =$ Yes and $x =$ Graduated. Then:

- $P(c|x) = P(\text{Yes} | \text{Graduated})$ means posterior probability of data.
- $p(c) = P(\text{Yes})$ means class prior probability of data.
- $p(x|c) = p(\text{Graduated} | \text{Yes})$ is the likelihood which is the probability of predictor given class in data.
- $p(x) = P(\text{Graduated})$ is the prior probability of the predictor of data.

So the final formula for our example would look like this:

$$P(\text{Yes} | \text{Graduated}) = P(\text{Graduated} | \text{Yes}) * p(\text{Yes})/P(\text{Graduated})$$

Now we find value for each part of the formula
Here:

- For $P(\text{Graduated} | \text{Yes})$, i.e., graduates with loan status Yes, which is $2/3 = 0.67$ since out of three Yes *Loan_status* only two are for graduate
- For $P(\text{Graduated})$, i.e., probability of graduates is $5/8 = 0.625$ since there are only five graduates out of eight data points
- Finally for $P(\text{Yes})$, i.e., probability of Yes is $3/8 = 0.375$ since there are only three yes out of eight *Loan_status*

Table 7.1 Loan status for education field

Education	Loan_status
Graduate	Yes
Graduate	No
Not graduate	Yes
Graduate	No
Not graduate	No
Graduate	Yes
Not graduate	No
Graduate	No

Now we substitute the values into our formula:

$$P(\text{Yes} \mid \text{Graduated}) = 0.67 * 0.375/0.625 = 0.402$$

Similarly, we will calculate posterior probability for No, i.e., chances of not getting loan approved. Formula for this case would be:

$P(\text{No} \mid \text{Graduated}) = p(\text{Graduated} \mid \text{No})^* \, p(\text{No})/P(\text{Graduated})$

- For $P(\text{Graduated} \mid \text{No}) =$ i.e., graduates with loan status No, which is $3/5 = 0.6$ since out of five No Loan_status only three are for graduate
- For $p(\text{No}) = 5/8 = 0.625$ since there are five No Status out eight Loan_status
- for $P(\text{Graduated}) = 5/8 = 0.625$

Now we substitute these values into our formula:

$$P(\text{No} \mid \text{Graduated}) = 0.6 * 0.625/0.625 = 0.6$$

So, by naive Bayes method of probability computations, since chance of not getting loan $P(\text{No}|\text{Graduated}) = 0.6$ is greater than the chance of getting loan $P(\text{Yes}|\text{Graduated}) = 0.402$, we predict the customer will not get a loan approval.

Having understood the application of naive Bayes method, let us look at the advantages and drawbacks of this theorem.

Advantages

The advantages of naive Bayes algorithm may be listed as follows:

- It is easy to implement.
- It is fast in training.
- Also performs well in multi-class prediction.
- If an assumption of independence holds, a naive Bayes classifier performs better compared to other models.
- As compared to other classifiers like logistic regression, it requires less training data.
- It performs well in case of categorical input variables compared to numerical variables. The discrete features are categorically distributed, while for numerical variables, we need to assume a normal distribution (bell curve), which is a strong assumption.

Disadvantages

Some drawbacks of naive Bayes classifier are:

- Zero frequency—if a certain categorical variable exhibits a new category in a test dataset which was not observed during training, the model assigns zero probability to such a variable, resulting in an inappropriate prediction. This is known as zero frequency. We use some smoothing techniques like Laplace estimation to solve this problem.
- Features independence —in real life, it is almost impossible to get a set of predictors which are completely independent.
- Bad estimator—some call naive Bayes as a bad estimator (quick and dirty) and suggest not to take its predictions seriously. It is good for quick and rough estimations only.
- As the hypothesis function is so simple, it fails to represent many complex situations and usually exhibits low variance to the target class.
- Because of the simplicity of the hypothesis, there is no over-fitting during training, and thus there is no way to validate if the trained model is properly generalized to predict unseen data.

Having considered the advantages and disadvantages, we will now look at how to improve the model's performance.

Improving Performance

Here are a few tips to improve the performance of naive Bayes classifiers.

- If a numeric feature does not have a normal distribution, use transformation or some other method to normalize it.
- If the test data set has zero frequency issue, apply Laplace correction or some other smoothing technique.
- Remove correlated features. Highly correlated features get voted twice and can lead to over-importance.
- The classifier has only a few options for parameter tuning, so focus more on data preprocessing and feature selection.
- As the classifier exhibits low variance, some improvement techniques like ensembling, bagging, and boosting will not help; the general purpose of these techniques is to reduce variance.

Now, as you have studied the naive Bayes model, let us look at its various types.

Naive Bayes Types

Depending on the data distribution, we use different implementations of naive Bayes classifiers. The scikit-learn and other libraries provide ready-to-use implementations of these classifiers. I will show you a few implementations.

Multinomial Naive Bayes

When your data has a multinomial distribution, you use the *MultinomialNB* classifier from sklearn. For example, if your application is to classify a document into categories like sports, politics, technology, etc., you would use word count vectors as features where, for each class, the probability of its occurrence in the document is prior computed. The implementation requires you to define the value of the smoothing parameter α (alpha), which is set to 1 for Laplace smoothing and less than one for Lidstone smoothing. The smoothing takes care of features not present in the learning samples and prevents zero probabilities for such values found in the test dataset. Movie ratings will be another application where this implementation is used. Note that the ratings would range from say 1 through 5, which is based on the frequency distribution of certain keywords in the document. This is an example of a multi-class classification.

Bernoulli Naive Bayes

You use this implementation when your predictors are Boolean, that is to say, you assume a multivariate Bernoulli distribution. There are multiple features, and we assume each one to be a binary-valued (Bernoulli, Boolean) variable. In multinational, we used the word count; in Bernoulli you use the word occurrence vector. We observed that this variant works better on shorter documents. You may still evaluate both models for acceptable accuracy scores.

Gaussian Naive Bayes

When your data distribution is normal, you use *GaussianNB* implementation of sklearn. To cite an example, consider the iris dataset of UCI repository, which probably every ML has used in his early days of learning. The dataset is about flowers with various features such as sepal width and length, petal width and length, and so on. All these features have a continuous distribution of data, so we would use this *GaussianNB* implementation for classification of flowers into different categories.

Complement Naive Bayes

What if you have an imbalance dataset? Here, the *CompleteNB* implementation would produce better results on text classification tasks as compared to multinomial NB implementation. We base the implementation on complement naive Bayes (CNB) algorithm. It uses statistics from the complement of each class to compute the model's weights.

Categorical Naive Bayes

If the data is categorically distributed, you use *CategoricalNB* implementation. Here, we assume each feature has its own categorical distribution. Encode each feature using *OrdinalEncoder* of sklearn to represent it as a set of numbers in the range 0 through $n_i - 1$ where n_i is the number of categories of feature i.

After discussing the various implementations of naive Bayes theorem, I shall discuss one more important aspect, and that is how to fit the model for an extremely large dataset.

Model Fitting for Huge Datasets

Consider the development of a model used for news categorization. Each news comprises hundreds of words and datasets, for millions of such news can be easily got or created. Loading such a dataset into memory for model fitting may not be a workable option. For such situations, sklearn provides a version of *fit* (model fitting) method called *partial_fit* that can be used incrementally on the dataset. We need to pass the list of expected class labels to the first call of this *partial_fit* method.

Having discussed the various aspects of naive Bayes classifier, I will now show you a practical example of training such a classifier.

Project

To demonstrate the development of naive Bayes classifier, I will use the News Aggregator dataset published in UCI Machine Learning Repository. The dataset consists of more than 400 thousand news stories collected by web aggregation. All these data points are labeled as one of the four categories—business (*b*), science and technology (*t*), entertainment (*e*), and health (*m*). We will train the naive Bayes classifier on this dataset.

Preparing Dataset

The original dataset contains many columns. I will use only the TITLE column to classify the news item. The CATEGORY column obviously is our target. I extracted these two columns and have put the resulting csv file on the book's repository. You can upload this file into your project using *wget* command. The first few records of the loaded dataset are shown in Fig. 7.2.

I did some preliminary text preprocessing on the *TITLE* column and copied the result into another column that I called as *TEXT*.

```
news['TEXT'] = [normalize_text(s)
                       for s in news['TITLE']]
```

Then, I tokenize the *TEXT* column using the built-in *CountVectorizer* function.

```
vectorizer = CountVectorizer()
x = vectorizer.fit_transform(news['TEXT'])
```

The target column was encoded using the following code:

```
encoder = LabelEncoder()
y = encoder.fit_transform(news['CATEGORY'])
```

Now, we are ready for model building. Just to understand the frequency or importance of the words created in our vocabulary, I used word cloud for some visualization.

At this point, our dataset is fully preprocessed and ready for machine learning. First, we will visualize the dataset to understand it better.

	TITLE	CATEGORY
0	Fed official says weak data caused by weather,...	b
1	Fed's Charles Plosser sees high bar for change...	b
2	US open: Stocks fall after Fed official hints ...	b
3	Fed risks falling 'behind the curve', Charles ...	b
4	Fed's Plosser: Nasty Weather Has Curbed Job Gr...	b

Fig. 7.2 Sample dataset

Data Visualization

We will use word cloud to visualize the data. It produces a graphical output where the most frequent words are shown in larger sizes. Figure 7.3 shows the output in my test:

Model Building

After splitting the dataset into training and testing, we apply the multinomial naive Bayes algorithm using the following code:

```
mnb = MultinomialNB()
mnb.fit(x_train, y_train)
```

We check the model's accuracy by calling its *score* method:

```
mnb.score(x_test, y_test)
```

```
0.9274773921689314
```

The output indicates that the model is giving 92.7% accuracy on the test data. Next, we will try the Bernoulli model.

```
bnb=BernoulliNB()
bnb.fit(x_train,y_train)
bnb.score(x_test, y_test)
```

Fig. 7.3 Determining word importance

The output in this case is:

```
0.9273471900004735
```

As you see, both models have given almost equal accuracy.

Inferring on Unseen Data

We will now create four title messages and study the inference made by our trained model on these messages. We use the following code:

```
text_message=[
    "New 'Game of Thrones' Season 4 Trailer: ",
    "I Will Answer Injustice with Justice!",
    "Bitcoin exchange files for bankruptcy",
    "ECB's Noyer: Low inflation may hamper adjustment"]

msg = vectorizer.transform(text_message)
print("prediction from model: {}".format(
                                    mnb.predict(msg)))
```

The output is:

```
prediction from model: [1 1 0 0]
```

So, the model has predicted that the first news item belongs to category 1 and the rest two belong to category 0.

The full source of this project is available in the book's repository.

Summary

In this chapter, you learn about the naive Bayes algorithm and its applications in real-life situations. A machine learning model based on this algorithm helps in making quick predictions on a high-dimensional dataset. This is probably the simplest and yet the most efficient algorithm for classification. It trains quickly and is thus very useful in real-time analysis of data. We base it on naive Bayes theorem and make a "naive" assumption that all features in the dataset are independent of each other. The sklearn library provides several implementations based on the data distribution.

In the next chapter, you will learn the support vector machines (SVM) algorithm.

Chapter 8
Support Vector Machines

A Supervised Learning Algorithm for Classification and Regression

Support vector machine, also called SVM in short, is one of the most frequently used algorithms for classification; it can also be used for solving regression problems.

In a Nutshell

SVM is a powerful, flexible supervised learning algorithm most commonly used for classification. It finds an optimal hyperplane to divide the dataset into two or more classes (groups). Once a hyperplane is determined, it would be easier for you to place an unseen data point into one of these regions.

To understand how the hyperplane is determined, I will explain the working of SVM.

SVM Working

Let us begin with a trivial case of separating the data between two classes. Figure 8.1 shows three hyperplanes separating a set of data points into two different classes.

The SVM algorithm will select the hyperplane marked with maximum margin as the separating boundary. Let me define a few technical terms used in SVM.

- Hyperplane—It is a decision boundary or a plane that helps classify data points. Depending on which side of the hyperplane a point lies, its class is determined. For a dataset that has only two classes, the hyperplane can be a simple line. For datasets having multiple classes, this can be a hyperplane in n-dimensional space.

P. Sarang, *Thinking Data Science*, The Springer Series in Applied Machine Learning, https://doi.org/10.1007/978-3-031-02363-7_8

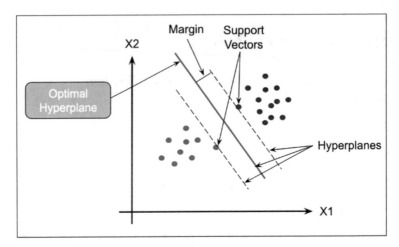

Fig. 8.1 Determining optimal hyperplane

- Support vectors—These are the data points that are closest to the hyperplane. They affect the positioning of the hyperplane and thus are called support vectors. Based on this, the algorithm for determining the hyperplane itself is called the support vector machine (SVM) algorithm.
- Margin—margin is the distance between the hyperplane and the support vectors. The algorithm always chooses the hyperplane that maximizes the margin.

Now, the margin can be hard or soft.

- Hard margin—when the training dataset is linearly separable, you can draw two parallel lines, as depicted in the figure below. We call this hard margin.
- When the dataset is not fully linearly separable, SVM allows some margin violation, as depicted in Fig. 8.2. Notice that some data points are on the wrong side of the hyperplane or between the margin and the hyperplane. We call this soft margin.

After the hyperplane is determined, we can classify an unseen data point as one of the known classes depending on which side of the hyperplane it lies.

Now, let us look at types of hyperplanes.

Hyperplane Types

The hyperplanes can be a simple straight line or can be a complex plane in n-dimensional space.

A linear and a nonlinear hyperplane in two dimensions can be seen in Fig. 8.3.

You can imagine in three or more dimensions; the hyperplane could be a complex surface. Some of such hyperplanes in multidimensional datasets are shown in Fig. 8.4.

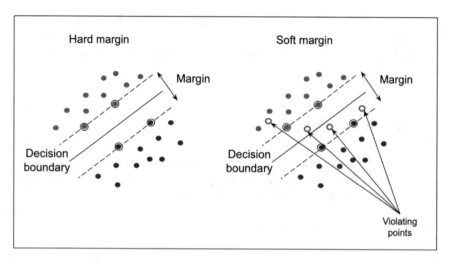

Fig. 8.2 Hard and soft margins

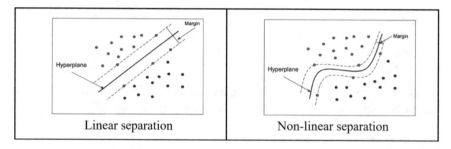

Fig. 8.3 Linear/nonlinear hyperplanes

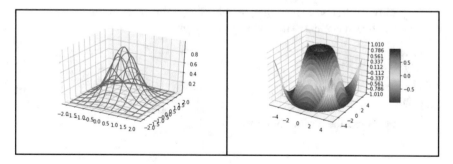

Fig. 8.4 Three-dimensional hyperplanes

You can understand that the computation of a hyperplane would be the most complex task that the algorithm has to perform. Fortunately, the implementers have done a great job of working with complex mathematics; and as a data scientist, our job is just to accept their results and use the hyperplane provided by them to classify an unseen data point.

Complex mathematics uses a "kernel" to create an optimal decision boundary. As a data scientist, without going into the kernel mathematics, let us look at the different kernels they have provided and their effects on our dataset. I will illustrate this with code snippets.

Kernel Effects

I will use the widely used *iris* dataset to show the effect of different kernels in classification of the petals. The dataset is provided in the sklearn library and is loaded into your project using the following two lines of code:

```
from sklearn import svm, datasets
iris = datasets.load_iris()
```

While applying the SVC (support vector classifier), we will specify which kernel to use. There are four kernel types defined in the sklearn library.

- Linear
- Polynomial
- Radial basis function
- Sigmoid

The sklearn library provides the implementation of a support vector classifier algorithm in *SVC* class. You call it using the following two lines of code:

```
svc = svm.SVC(kernel='linear')
svc.fit(X, y)
```

To specify a different kernel type, simply change the value of the *kernel* parameter in the above statement. The possible values are:

- linear
- poly
- rbf
- sigmoid

I will now describe each one of these kernels.

Linear Kernel

This is used when your dataset is linearly separable. Mathematically, it is represented as:

$$K(x, y) = sum(x * y)$$

where x and y are two vectors.

You will use this kernel for its quick training on many features. The training is faster than other kernels as it uses only the C regularization parameter for optimization, while the others use *gamma* parameter. I will explain the regularization parameters in the next section "Parameter tuning".

Polynomial Kernel

This is a generalized form of the linear kernel that will classify the nonlinear datasets. Mathematically, it is represented as:

$$K(x, y) = (x^T * y + c)d$$

where x and y are the two vectors, constant c allows trade-off for higher and lower dimension terms, and d is the order of the kernel.

Radial Basis Function

This is also known as the Gaussian kernel and is most widely used. It has the ability to map the input data into indefinite high dimensional spaces. Mathematically, it is represented as:

$$K(x, y) = exp\left(-gamma * sum\left(x - y^2\right)\right)$$

where x and y are the two vectors and *gamma* is a tuning parameter with a range 0 to 1.

Sigmoid

This is equivalent to the sigmoid activation function used in artificial neural networks. It outputs only 0 and 1 values and thus is used only for binary classification. Mathematically it is represented as:

$$K(x, y) = tanh\left(\gamma.x^T y + r\right)$$

The effect of these four different kernels on our dataset can be seen in Fig. 8.5.

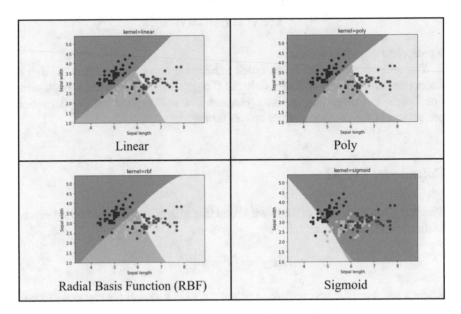

Fig. 8.5 Hyperplanes for different kernels

You can clearly see that the sigmoid kernel fails to classify the multi-class dataset that we have here. The full source of this project is available in the book's repository.

Guidelines on Kernel Selection

As a data scientist, you need to decide on which kernel to apply on your dataset. I will provide you with a few guidelines that will help you make this decision.

As you have seen, the three kernels—linear, polynomial, and radial basis functions, differ in their mathematical approach while creating hyperplanes. The linear and polynomial kernels have less training times at the cost of accuracy. The radial basis function kernel is more accurate but requires larger training time. So, the decision on which kernel to use would depend on the complexity of your dataset and the desired accuracy. If by examination or with the domain knowledge you know that your dataset is linearly separable, use a linear kernel as it requires low training time. If you know that your dataset is nonlinear, use the polynomial for quicker training, but less accuracy than *rbf*. Use *rbf* on your nonlinear datasets for better accuracy at the cost of higher training times.

The *SVC* function call besides the kernel parameter also takes a few parameters with some default values. When you apply *SVC* on your dataset, you may also like to know the effect of these parameters to fine-tune the model. I will now discuss the few important parameters in the *SVC* call.

Parameter Tuning

I will discuss the four parameters:

- C
- degree
- gamma
- decision_function_shape

The C Parameter

This is a float type parameter having a default value of 1.0. We use the value of this in regularization, which is a technique used to reduce the errors and avoid over-fitting. The regularization strength is inversely proportional to the value of C. The value of C must be strictly positive. The penalty is squared $l2$. The following code snippet shows the effect of various C values on our dataset.

```
C_values = [0.1, 1, 10, 100, 1000, 10000]
for ci in C_values:
    svc = svm.SVC(kernel='rbf', C=ci)
    svc.fit(X, y)
    plotSVC('C = ' + str(ci))
```

The hyperplanes generated for various C values can be seen in Fig. 8.6.

Observe how the contours and thereby the regions change with the increasing value of C parameter.

The Degree Parameter

This is of int type having a default value of 3. We apply this parameter only on the polynomial kernel and decide the degree of the polynomial used in the computation. If you specify this parameter for other kernels, it will ignore its value. If you use a higher value for this parameter, it will increase your training times. The following code snippet illustrates the effect of different values on the degree.

```
D = [0, 1, 2, 3, 4, 5]
for degree in D:

svc = svm.SVC(kernel='poly', degree=degree)
svc.fit(X, y)
plotSVC('degree = ' + str(degree))
```

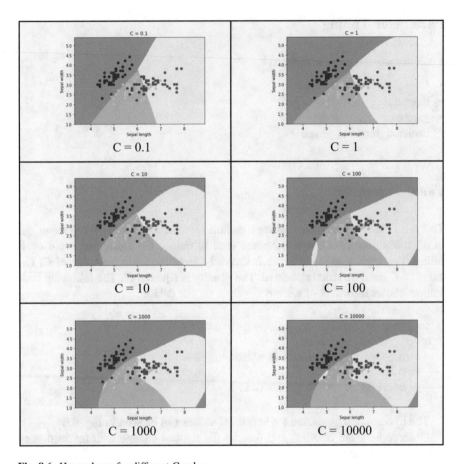

Fig. 8.6 Hyperplanes for different C values

```
svc_default = SVC()
svc_default.fit(X_train,y_train)
```

The hyperplanes generated for various degrees can be seen in Fig. 8.7.

Once again observe how the hyperplane changes with the degree value. Note that for the degree equal to zero, we do not get any hyperplane.

The Gamma Parameter

This is of float type, and we can specify its value as *scale* or *auto*. The default value is *scale*. This parameter determines how much curvature we want in a decision boundary. For the default scale value, it sets the gamma to 1 / (n_features * X.var()), while for auto, it sets gamma to 1 / n_features. Figure 8.8 shows the effect of gamma.

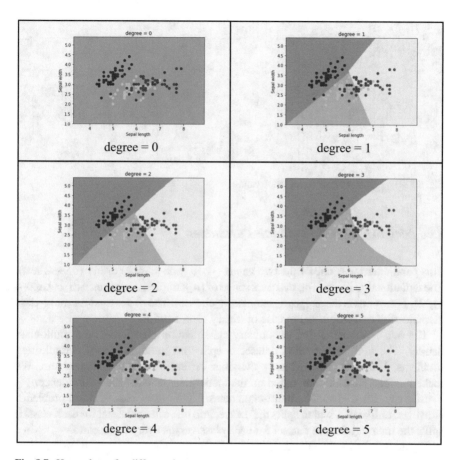

Fig. 8.7 Hyperplanes for different degrees

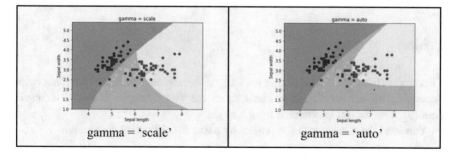

Fig. 8.8 Hyperplanes for different gamma

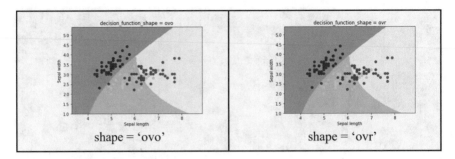

Fig. 8.9 Hyperplanes for different shape values

The decision_function_shape Parameter

This parameter takes one of the two values—*ovo* (one-vs-one) or *ovr* (one-vs-rest). The default value is *ovr*. The ovo uses shape of (n_samples, n_classes * (n_classes— 1) / 2). The *ovr* uses the shape (n_samples, n_classes). The *ovo* is used for multi-class classification and is ignored in case of binary classification.

The ovr is a method that uses binary classification algorithms on a multi-class dataset. The entire multi-class dataset is split into multiple binary classification problems. We then train a binary classifier on each such divided problem. We make the final classification based on the model that gave the most confidence.

Like ovr, the ovo splits a multi-class dataset into binary classification problems. However, unlike ovr, which splits the dataset into one binary dataset for each class, it splits the dataset into one dataset for each class versus every other class.

Figure 8.9 shows the output for the two different values of the decision function.

The full source code for the above simulations is available in the book's repository.

I will now present a practical use case of the SVM algorithm.

Project

This project uses an SVM classifier to distinguish between a male and a female voice depending on the acoustic characteristics of the voice sample. It has 19 acoustic features. The dataset is obtained from Kaggle.

You load the dataset into the project by using the following command:

```
!wget 'https://raw.githubusercontent.com/profsarang/
ThinkingDataScience/main/data/voice.csv'
```

Use the *info* method to see what it contains. The output is shown in Fig. 8.10.

Fig. 8.10 Dataset
information

```
df.info()

<class 'pandas.core.frame.DataFrame'>
RangeIndex: 3168 entries, 0 to 3167
Data columns (total 21 columns):
 #   Column    Non-Null Count   Dtype
---  ------    --------------   -----
 0   meanfreq  3168 non-null    float64
 1   sd        3168 non-null    float64
 2   median    3168 non-null    float64
 3   Q25       3168 non-null    float64
 4   Q75       3168 non-null    float64
 5   IQR       3168 non-null    float64
 6   skew      3168 non-null    float64
 7   kurt      3168 non-null    float64
 8   sp.ent    3168 non-null    float64
 9   sfm       3168 non-null    float64
 10  mode      3168 non-null    float64
 11  centroid  3168 non-null    float64
 12  meanfun   3168 non-null    float64
 13  minfun    3168 non-null    float64
 14  maxfun    3168 non-null    float64
 15  meandom   3168 non-null    float64
 16  mindom    3168 non-null    float64
 17  maxdom    3168 non-null    float64
 18  dfrange   3168 non-null    float64
 19  modindx   3168 non-null    float64
 20  label     3168 non-null    object
dtypes: float64(20), object(1)
memory usage: 519.9+ KB
```

After some preprocessing and creating training/testing datasets, we apply the SVM classifier algorithm with default parameter values as follows:

```
svc_default = SVC()
svc_default.fit(X_train,y_train)
```

After the model is trained, you can check its accuracy on the test dataset.

```
y_pred = svc_default.predict(X_test)
print("Accuracy Score:")
print(metrics.accuracy_score(y_test,y_pred))
```

In my run, I got an accuracy score of 0.98 percent, which is extremely good. To check the effect of different kernels on the accuracy score, I ran tests with different kernel values. We specify the kernel value as a parameter to *SVC*, as shown here:

Table 8.1 Accuracy scores for various kernels

'linear'	'rbf'	'poly'
0.9779179810725552	0.9763406940063092	0.9589905362776026

```
svc_linear=SVC(kernel='linear')
svc_RBF= SVC(kernel='rbf')
svc_poly=SVC(kernel='poly')
```

Table 8.1 gives the accuracy scores for the three kernels.

As you observe, the accuracy score remains more or fewer the same in all three cases—this is mainly because the dataset is perfectly balanced, having 50/50 data points for male and female categories.

The full source code for this project is available in the book's repository.

Now, I will give you a few advantages and disadvantages of the SVC.

Advantages and Disadvantages

Here are some advantages of SVM:

- It handles nonlinear data efficiently using kernel tricks.
- A minor change in the dataset does not affect the hyperplane, and thus the model is considered quite stable.
- We can use the algorithm for both classification (SVC) and regression (SVR).
- The algorithm has good generalization capabilities due to $l2$ regularization; this prevents the model from over-fitting.

Here are the disadvantages:

- Choosing an appropriate kernel function is a hard task. Especially, if you use high dimensionality for a kernel, you might generate too many support vectors that ultimately result in dramatically low training speeds.
- As the algorithm is quite complex, the memory requirements are very high. This is mainly because the algorithm has to store all support vectors in memory, and this number grows exponentially with the size of the training dataset.
- The algorithm requires all features to be scaled to a common scale before its application.
- The training times are usually very long, especially on large datasets.
- The output of this algorithm, which is a hyperplane, is generally difficult to understand and interpret, unlike decision trees which can be visualized and interpreted easily by human beings.

Summary

The SVM algorithm finds an optimal hyperplane to divide the dataset into several classes. The hyperplanes need not be a simple straight line, but it can be a complex shape depending on the number of features and the target values. The algorithm uses kernel tricks to find the optimal hyperplane. You studied the various kernel types and the effect of other parameters in determining the best hyperplane. SVM is a very popular and widely used algorithm for classification and can also be used for regression tasks.

In the next chapter, I will introduce you to clustering—a challenging task for data scientists.

Part II
Clustering: Overview

Clustering is an important aspect of data analytics and the most important unsupervised learning. It aims to divide a dataset into subsets. Objects in the same subset are like each other regarding a certain similarity measure. Objects in different clusters are dissimilar in the same measure. You use clustering for understanding the data or as a first step in further operations like indexing and data compression. The need for understanding the data comes in for pattern recognition, data mining, information retrieval, and so on.

The clustering process itself broadly follows these steps:

- Extract and select the most representative features.
- Design a new or select a known clustering algorithm.
- Evaluate the results to judge algorithm's validity.
- Explain the results to your customer. If the customer is satisfied with your explanation, your job is done.

As the notion of a "cluster" is not precisely defined, the researchers developed many clustering models and thus many clustering algorithms. Understanding these models is important in learning these algorithms. The hierarchical clustering builds models called connectivity-based on distance connectivity. The centroid models, like the K-means algorithm, form clusters by a single mean vector. The distribution models cluster data based on its statistical distribution. We use expectation-maximization (EM) algorithm to find clusters in multivariate distributions. Then there are graph-based models like CLIQUE. Some algorithms like DBSCAN provide just the grouping information for clustering the subgroups. The grid-based clustering, like STING, divides the dataset into subspaces. You even have gossip-style clustering, such as affinity propagation, that uses messaging between peers to form groups or clusters. You also have neural models like a self-organizing map which is an unsupervised neural network.

As a data scientist, you will need to apply one or more of these models to provide meaningful analytics to your customer. The clustering algorithms broadly fall under these categories:

- Centroid-based
- Connectivity-based
- Distribution-based
- Density-based
- Clustering huge datasets
- Gossip-style clustering
- Grid-based

In centroid-based clustering, we start with the assumption that the data would contain K clusters. Clusters are formed by solving the optimization problem defined as find K cluster centers and assign the objects to the nearest cluster center. A distance function measures the distance between the points to determine the nearest center. The optimization problem is NP-hard, and thus this technique can give you only approximate solutions. The K-means clustering falls under this category. The K-medoids, K-medians, and K-means++ are a few variations. The biggest drawback of this algorithm is that it requires you to estimate the number of clusters in advance. We also assume that the clusters have approximately the same size. As the algorithm optimizes cluster centers and not the borders, we reach incorrectly cut borders. The mean shift clustering is yet another algorithm that comes under this category. This algorithm discovers the clusters in a smooth density of data points by iteratively shifting the kernel to a higher density region until convergence.

In connectivity-based clustering, you connect objects to form clusters based on their distance. We assume that the nearby objects are more related to each other than the farther away objects. Some maximum distance decides a cluster. We can represent the entire structure as a dendrogram, and that is why this type of clustering is also called hierarchical clustering. As you traverse the dendrogram, the clusters merge with each other at certain distances. The clustering depends not just on the choice of distance function but also on the choice of linkage criterion. It can be a single, complete, or average linkage. The clustering can be agglomerative or divisive. In agglomerative, you start with a single data point and aggregate other points to it to form a cluster. In divisive, you consider the entire dataset as a single cluster and then divide it into partitions. From this description, you can understand that these methods do not produce a unique partitioning of data, rather they create a hierarchy from which the user has to choose appropriate clusters.

The distribution-based clustering is based on the statistical distribution models. Objects belonging to the same distribution form a cluster. A typical distribution is Gaussian—we assume the entire dataset to comprise clusters, each having a Gaussian distribution of points within it. The Gaussian mixture models is one such clustering technique. This type of clustering is good at capturing correlation and dependence between attributes. However, the assumption (Gaussian distribution) we base them on is rather strong and thus may not produce good clustering if the data points do not follow the expected distribution.

In density-based clustering, high density of data points in a certain area defines the cluster. A drop in density marks the cluster borders. Such algorithms would work nicely for distributions like Gaussian mixtures. The DBSCAN clustering algorithm falls under this category. Like linkage-based clustering, we group the points lying within a certain distance threshold with the additional condition that there are a minimum number of other objects within the same radius. As the cluster comprises all density-connected objects, it can cause creating clusters of any arbitrary shape. The complexity of this algorithm is fairly low, and the outputs are deterministic—discovering the same results on every run. OPTICS is yet another algorithm that lies in this category. It generalizes upon DBSCAN by removing the need to change the value for range parameter ε. It produces a visualization of linkages between the points; the user creates clusters of his choice using these visualizations. Mean shift clustering is another algorithm in this category wherein each data point is moved to the nearest dense area based on kernel density estimation. The mean shift can detect arbitrary-shaped clusters. It is usually slower than DBSCAN and does not produce good clustering on multidimensional datasets.

How do you cluster a really enormous dataset? You use the divide-n-conquer strategy. For really huge datasets, we divide it into smaller ones, keeping as much information as possible from the original in the splits. We will then cluster each compact unit using the other known clustering techniques. This is the way the BIRCH clustering algorithm works. CLARANS is another algorithm that we use for clustering large datasets. It partitions the large dataset with a randomized search into smaller units. It then applies the PAM (partitioning around medoids) algorithm on each subset, determines the clustering quality by measuring the dissimilarities, and repeats the entire sampling and clustering process until it satisfied you with the clustering output.

Then comes another category of messaging, which is gossip-based. We call it affinity propagation clustering, where we form the groups or clusters by gossiping—messaging between the peers.

All above algorithms are query-based, where the intermediate results of a query are not saved. You have to scan the entire dataset for each query. So, here comes another class of clustering, called grid-based. STING is one such type; another one is CLIQUE, which is considered both grid and density-based.

Table 1 provides a comprehensive summary of clustering algorithms.

The list given in Table 1 does not surely cover every known algorithm. As a data scientist, get at least a conceptual overview of these or at least few algorithms in each

Table 1 List of clustering algorithms

Category	Algorithms
Partition-based	K-means, K-medoids, PAM, CLARA, CLARANS
Hierarchy-based	BIRCH, ROCK
Density-based	DBSCAN, OPTICS, mean shift
Grid-based	STING, CLIQUE
Model-based	GMM, SOM
Graph theory-based	CLICK, MST
Distribution-based	GMM
Fuzzy theory-based	FCM, FCS, MM

category. Clustering is not a straightforward task for anybody, especially because of its unsupervised nature. There is no known result or you may not get a consensus on the results.

You now have an overview of several major clustering algorithms. The following chapters would give you a conceptual overview of these algorithms without going into the implementation or the mathematics behind it. After all, you only need to know the purpose of the algorithm and what kind of clustering problem it can address. This is a list of algorithms that you will study in the next few chapters:

- K-means
- K-medoids
- Mean shift
- Agglomerative
- Divisive
- Gaussian mixture model
- DBSCAN
- OPTICS
- BIRCH
- CLARANS
- Affinity propagation
- STING
- CLIQUE

So, let us begin with the first algorithm in our list, and that is the K-means algorithm (centroid-based clustering).

Chapter 9
Centroid-Based Clustering

Clustering Algorithms for Hard Clustering

In this chapter, you will study two important centroid-based clustering algorithms:

- *K*-means
- *K*-medoids

We begin with *K*-means algorithm, which is usually considered a starting point for estimating the number of clusters in your dataset—let it be small or big.

The *K*-Means Algorithm

In a Nutshell

When your dataset is of small to medium size and you want to find the number of clusters in it, use this algorithm. The algorithm finds the best value for centroids and groups all nearby objects into the respective clusters. The optimization is NP-hard and thus usually providing only approximate solutions. The algorithm provides hard clustering, meaning each data point belongs only to a single cluster. Figure 9.1 shows the difference between hard and soft clustering.

Let us now look at how *K*-means works.

How Does It Work?

K-means is an unsupervised learning algorithm that groups the unlabeled data points into different clusters. The algorithm creates the specified number (K) of clusters. While grouping, the algorithm determines the best value for the centroid by an iterative process. Figure 9.2 shows the three identified clusters in the original dataset after running the algorithm with the initial value of $K = 3$.

© The Author(s), under exclusive license to Springer Nature Switzerland AG 2023
P. Sarang, *Thinking Data Science*, The Springer Series in Applied Machine Learning,
https://doi.org/10.1007/978-3-031-02363-7_9

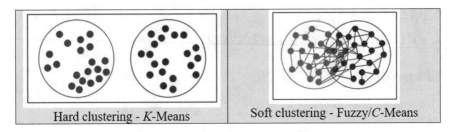

Fig. 9.1 Hard and soft clustering

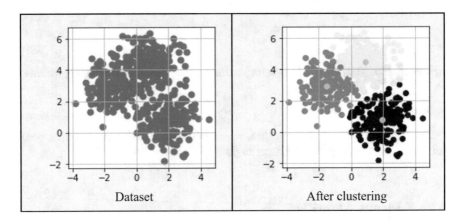

Fig. 9.2 Original and clustered datasets

This is how it works.

To begin with, you need to decide on the number of clusters you want to create in your dataset. This is the K value in the algorithm. It means that you need to estimate the number of clusters in the dataset before running the algorithm. This could be a hard task. Fortunately, we have few techniques for estimating this K value. I will explain those to you shortly.

Let us consider you decide on K equal to 3. Now, select three centroids at random. This is your cluster initialization. Measure the Euclidean distance between each data point and the centroid. Assign each point to the nearest centroid for cluster formations. Compute the mean of each cluster. This will be your new centroid to try out. Repeat the entire procedure with the new centroids to create a new set of clusters. Do this repeatedly until convergence, that is, no further changes or the maximum number of iterations decided by you have reached.

Does the algorithm stop here? No, you still need to consider the variance in each cluster. Your aim should be to choose centroids that minimize the sum-of-squares within each cluster. So, you start all over again with a new set of three random centroids. Do the same exercise to form another set of clusters. Repeat this process as many times as you want. Now, you have a set of observations, where each observation gives a set of three clusters. Select the observation that gives the three clusters with the lowest sum of variance. This is your final output.

I will now give you the algorithm.

K-*Means Algorithm*

These are the steps of the K-means algorithm:

1. Assign an estimated value (number of clusters) to K.
2. Randomly select K data points as centroids.
3. Repeat the following steps until convergence (no further changes observed) or reach the pre-specified number of iterations:

   (a) Compute the sum of the squared distances between all data points and the centroids.
   (b) Assign each data point to the closest centroid.
   (c) Compute the new centroids by taking the mean of all data points belonging to each cluster

Here is the optimization objective.

Objective Function

Given a set of n observations $(x_1, x_2, \ldots, x_n)$, where each observation is a d-dimensional real vector, we aim to partition all observations into k ($\leq n$) sets $\{S_1, S_2, \ldots, S_k\}$ so as to minimize the WCSS (within-cluster sum of squares).

Mathematically, the objective is expressed as:

$$\underset{S}{\text{argmin}} \sum_{i=1}^{k} \| X - \mu_i \|^2$$

where μ_i is the mean of points in S_i.

This is an NP-hard problem where the time complexity is $O(n^{dk+1})$, n being the number of observations, k the number of clusters, and d the dimension.

Note that you will need to repeat this process of finding clusters multiple times and then select the best set of clusters.

I will now give you the entire process workflow.

The Process Workflow

The entire process workflow can be diagrammatically represented as seen in Fig. 9.3.

Next, I will show you how to optimally select the initial K value.

Fig. 9.3 Workflow for
K-means clustering

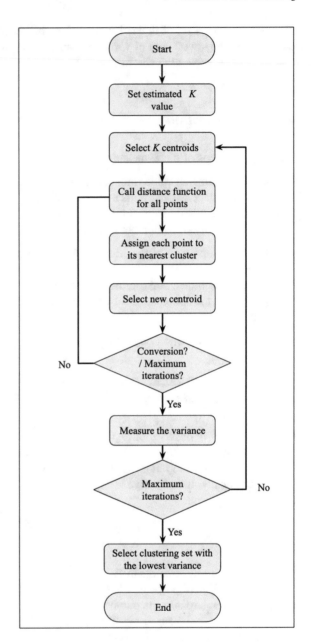

Selecting Optimal Clusters

When you look at the dataset that you want to divide into clusters, it is hard to
visualize how many clusters could be formed. We do have a few techniques
available to us for this purpose. We call these:

- Elbow
- Silhouette
- Gap statistics

Next, I will discuss these three methods.

Elbow Method

This is probably the most popular method that is based on the concept of WCSS (within-cluster sum of squares). As seen earlier, WCSS is mathematically expressed as:

$$WCSS = \sum_{i=1}^{k} \sum_{x \in s_i} \| x - \mu_i \|^2$$

where distance can be Euclidean distance as typically used in *k*-nearest neighbors (classification) algorithm.

We execute the algorithm for different K values, say in the range of 1 to 10, measuring WCSS for each. We then plot WCSS vs K as seen in Fig. 9.4.

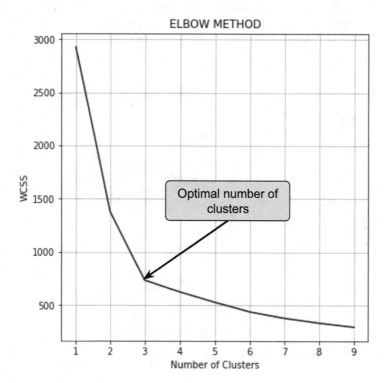

Fig. 9.4 Elbow method of determining optimal clusters

We can consider the point at which the curve flattens as the optimal value for K. As the curve looks like an Elbow, we call it the *elbow* method.

Average Silhouette Method

We base this method on measuring the cluster quality by determining how well each object lies within its cluster. To measure the quality, we define a silhouette coefficient:

$$\text{Silhouette coefficient } (s_i) = (x - y) / max \ (x, y)$$

where y is the mean intracluster distance, that is, the mean distance to the other data points in the same cluster. The x represents the mean nearest cluster distance, which is the mean distance to other data points of the next closest cluster. The value of s_i varies between -1 and 1. The value close to 1 shows that the data point is close enough to its cluster to be considered as a part of it. A value close to -1 shows that we assign the data point to the wrong cluster.

With this definition of silhouette coefficient, we design the algorithm on similar lines of the elbow method. We run the algorithm for different values of K, say in the range 1 to 10. For each K, we compute the average silhouette and then plot this score against K values. The peak of the curve, that is, the maximum silhouette score, decides the best value of K. Figure 9.5 shows such a plot.

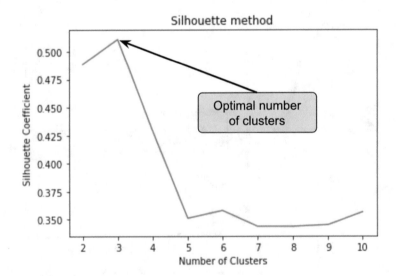

Fig. 9.5 Silhouette method of determining optimal clusters

The Gap Statistic Method

While the silhouette method measures the quality, the gap statistic measures how good is the clustering. Stanford researchers developed this method. It compares the *log* W_k (within-cluster sum of squares) with a null reference distribution of data. A null distribution is a distribution with no obvious clustering. The optimal *K* is the value for which this log falls the farthest below the reference curve. Mathematically, the gap is expressed as:

$$Gap_n(k) = E_n^*\{Log\ W_k\} - log\ W_{k^*}$$

You generate the reference dataset by sampling uniformly within the original dataset's bounding rectangle. In Fig. 9.6 we observe the gap is maximum for *K* equals three. This will be the optimal value for clustering.

Through the above three methods, you will determine the value of *K* to be used in the *K*-means algorithm. However, for certain types of data distributions, the *K*-means algorithm will not produce satisfactory clustering.

Limitations of **K-Means Clustering**

I have shown two distributions in Fig. 9.7, where the *K*-means algorithm will not produce satisfactory results.

The one on the left shows a distribution having three clusters, each having a different size. The one on the right again depicts three clusters. Here, the sizes are

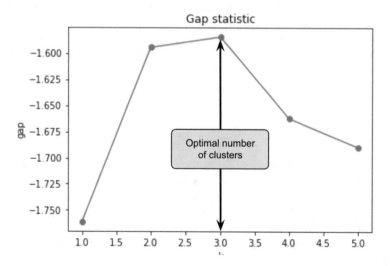

Fig. 9.6 Gap statistic method of determining optimal clusters

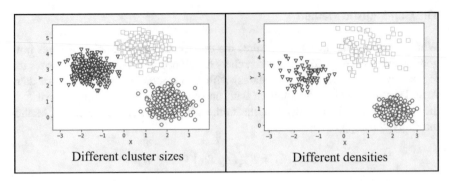

Fig. 9.7 Clusters having different sizes and densities

almost the same, but the density of points in each one is different. In both situations, the K-means algorithm will fail to form appropriate clusters. The solution in such situations could be to try out a larger K value.

Another issue in K-means clustering is the random initialization of the centroids, as each run would generate a unique set of clusters. To solve this problem, another algorithm called *K-means++* was introduced that chooses the initial centroids.

Applications

The K-means algorithm has been successfully applied in many areas, such as computer vision, document clustering, market segmentation, and so on. We often use it to get a meaningful intuition of the structure of the large datasets. We then apply for more advanced algorithms which are not so resource-hungry to do further clustering. To use K-means on large datasets, we use other algorithms, such as Lloyd's for using heuristics.

With this conceptual understanding of the K-means algorithm, I will now discuss its implementation as provided in sklearn library.

Implementation

The sklearn library provides the implementation of the K-means algorithm in the *sklearn.cluster.KMeans* class. Using this class is quite trivial—you just need to specify the K value during its instantiation; the rest of the parameters can take their default values. The class also supports *K-means++* whereby you can specify your desired centroid at the start.

To illustrate the implementation, I have created a small project that also shows the three methods for obtaining optimal K.

Project

We first create a dataset of 500 random samples by calling the *make_blobs* method.

```
X, Y = make_blobs(n_samples=500, random_state=0,
                  centers=3, n_features=2,
                  cluster_std=0.9)
```

The generated data looks like the one shown in Fig. 9.8.
We now cluster the dataset by calling the library provided *KMeans* function.

```
kmeans = KMeans(n_clusters=3)
kmeans.fit(X)
```

After the clustering, we will do the prediction on the same dataset.

```
km_pred = kmeans.predict(X)
```

Figure 9.9 shows the clustering result.

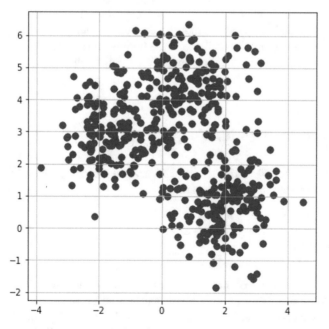

Fig. 9.8 Random dataset

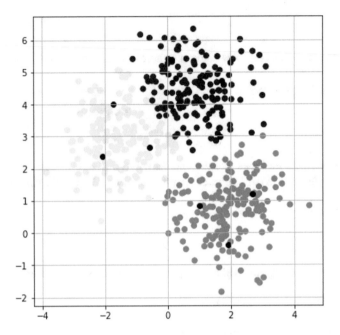

Fig. 9.9 Clustered dataset

You can see how easy it is to cluster a dataset. The only requirement is a good estimation of the number of clusters. In the project, I have also provided the implementation code for elbow, silhouette coefficient, and gap statistic methods.

The full project source is available in the book's download.

I will now discuss another important algorithm in this category, and that is *K-medoids*.

The *K*-Medoids Algorithm

Like *K*-Means, the *K*-Medoids is an unsupervised clustering algorithm that overcomes some shortcomings of *K*-means.

In a Nutshell

In *K*-means clustering, you have seen that we selected the average of a cluster's points as its center. This average point may not be an actual data point in your dataset. In *K*-medoids, we always pick up the real data point as the cluster center. This center is called medoid or an exemplar. It is the most centrally located data point in the cluster, with the minimum sum of distances to other points. There is another clustering algorithm in this category and that is *K*-medians, in which the median becomes the cluster center.

Why **K-Medoids?**

The *K*-means algorithm is sensitive to outliers, while the *K*-medoids algorithm is more robust to noises and outliers. To establish this claim, I created a random dataset, computed *K*-means and *K*-medoids on it; then added a few outliers to the dataset and re-calculated the two values. The centroid shift in *K*-means was much larger than the one observed for *K*-Medoids. Figure 9.10 shows the observations.

The project source for this experiment is available in the book's repository.

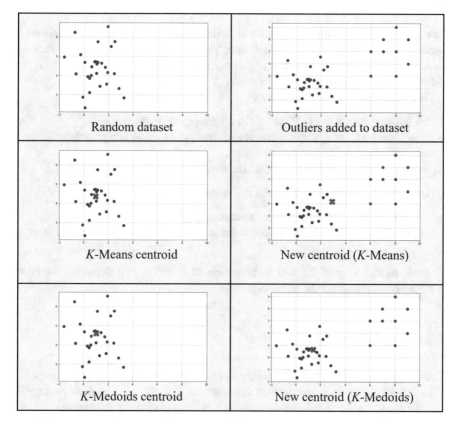

Fig. 9.10 Effect of outliers on *K*-means and *K*-medoids

As you see in the figure, a few outliers have affected the cluster centroid substantially for K-means. We clearly observed that K-medoid has produced better clustering than K-means. The reason behind this robustness towards noise and outliers is that K-Medoids minimize a sum of pair-wise dissimilarities rather than a sum of squared Euclidean distances, as with K-means. Both K-medoids and K-means are partitioned algorithms that break the dataset into groups to minimize the distance between the cluster center and the other points labeled to be in the cluster.

Another reason for choosing K-medoid over K-means is that we can use this algorithm with arbitrary dissimilarity measures, unlike K-means, where we use Euclidean distance for efficient solutions. In K-medoids, you may use Manhattan, Euclidean, or even cosine measures. I will show you the clustering done by these three different dissimilarity measures in the project that follows.

As there are many heuristic solutions, K-medoids is an NP-hard problem. Let us now look at the algorithm itself.

Algorithm

The K-medoids algorithm, also known as PAM (partitioning around medoids) uses a greedy search to find the optimal solution. The greedy search works faster than an exhaustive search. The algorithm requires you to input the value of K, which we estimate using methods such as silhouette.

We write the algorithm as follows:

1. Initialize: greedily select K out of the n data points as the medoids.
2. Associate each data point to the closest medoid by using any arbitrary distance metrics.
3. Perform the following swap, while the cost of the configuration decreases:

   (a) Let m be the medoid and o be the non-medoid data point.

      (i) Swap m and o, compute the cost change.
      (ii) If the cost change is the current best, remember this m and o as a combination.

4. Perform the swap of the best combination of m and o, if it decreases the cost. Otherwise, terminate the algorithm.

Merits and Demerits

The PAM algorithm provides flexibility in selection of similarity measure methods. It is also robust to noise and outliers as compared to K-means. The disadvantage is that we cannot apply it to large datasets as it is computationally expensive.

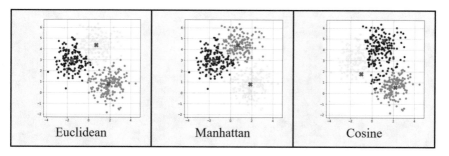

Fig. 9.11 Clustering with different metrics

Implementation

The sklearn library provides the implementation of the *K*-medoids algorithm in the *sklearn_extra.cluster.KMedoids* class. Using this class is quite trivial like *K*-means—you just need to specify the *K* value during its instantiation, the rest of the parameters can take their default values. The following code snippet shows you how to call the implementation:

```
model = KMedoids(metric="euclidean", n_clusters=1)
model.fit(X)
kmed_pred = model.predict(X)
centers = model.cluster_centers_
```

As with *K*-means I created a random dataset and applied *K*-medoids with different metrics. We show the results in Fig. 9.11.

The entire project source is available in the book's repository.

Summary

In this chapter, you studied two centroid-based algorithms—*K*-means and *K*-medoids. The algorithms required you to prior estimate the number of clusters in your dataset. This could be a tough job. You studied three different methods—elbow, silhouette, and gap statistic that help you in estimating the *K* value. The *K*-means algorithm considers the mean of the cluster points as the cluster centroid. This may not be an actual point in the dataset. The *K*-medoid algorithm takes the actual data point as its centroid and then decides on the neighboring points to be included in the cluster. The *K*-medoid algorithm is robust to outliers.

In the next chapter, you will study connectivity-based clustering algorithms.

Chapter 10
Connectivity-Based Clustering

Clustering Built on a Tree-Type Structure

In the previous chapter, you studied centroid-based clustering. In this chapter, you will learn the technique that uses a tree-type structure to create clusters. You will study two algorithms: agglomerative and divisive.

Agglomerative Clustering

In a Nutshell

Agglomerative clustering (AGNES) clusters your dataset by building a hierarchical tree-type structure. It uses a bottom-up approach while building the tree. It is more informative than the unstructured set of flat clusters created by K-means clustering. Unlike K-means, where you need to specify the value of K in the beginning, in AGNES, you do not have to specify the number of clusters. The bottom-up approach treats each data point as a single cluster at the outset. The algorithm then successively agglomerates pairs of clusters. The process continues until we merge all data points into a single cluster. We see this in Fig. 10.1. The dendrogram on the right shows the tree structure.

The Working

Implementing AGNES is trivial. These are the steps in the algorithm:

1. Let each data point represent a cluster.
2. Compute the proximity matrix of other data points using some measure such as Euclidean or Manhattan distance.

© The Author(s), under exclusive license to Springer Nature Switzerland AG 2023   185
P. Sarang, *Thinking Data Science*, The Springer Series in Applied Machine Learning,
https://doi.org/10.1007/978-3-031-02363-7_10

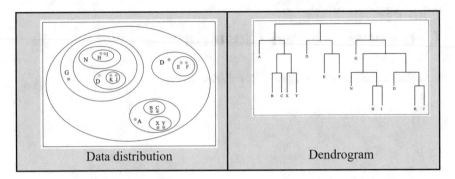

Fig. 10.1 Data distribution and dendrogram

3. Merge two closest clusters and update proximity matrix.
4. Repeat step 3 until a single cluster remains at the end.

From this working, you can easily understand how the hierarchical clustering is formed with a bounding single cluster on the outside.

While updating the proximity matrix in the above step 3, we consider three different linkages between the clusters. These are:

- Single
- Complete
- Average

I will now describe these three linkages.

Single Linkage

In single linkage, the distance between two clusters is the minimum value of all pairwise distances between the elements of the two clusters C_1 and C_2. This is also referred to as minimum linkage.

The minimum distance is calculated using the following formula:

$$Minimum\ distance : dist_{min}\left(C_i C_j\right) = \frac{min}{p \in C_1, p' \in C_j}\left\{|\,p - p'\,|\right\}$$

where $|\,p - p'\,|$ is the distance between two objects or points—C_i and C_j.

The formula may also be written as follows:

$$Dist(C_1 C_2) = Min(Dist(i, j)), i \in C_1 \text{ and } j \in C_2$$

Complete Linkage

In this type of linkage, the distance between two clusters is the maximum of all pair-wise distances between the elements of the two clusters C_1 and C_2. This is also referred to as maximum linkage.

The minimum distance is calculated using the following formula:

$$Maximum\ distance : dist_{max}(C_i,C_j) = \frac{max}{p \in C_1, p' \in C_j}\{|p - p'|\}$$

where $|p - p'|$ is the distance between two objects or points in C_i and C_j.

The formula may also be written as follows:

$$Dist(C_1,C_2) = Max(Dist(i,j)), i \in C_1 \text{ and } j \in C_2$$

Average Linkage

In this type of linkage, we consider the average distance, rather than the min or the max. This is also called UPGMA—unweighted pair group mean averaging.

The average is computed using the following formula:

$$Average\ distance : dist_{avg}(C_i,C_j) = \frac{1}{n_i n_j} \sum_{p \in C_1, p' \in C_j} |p - p'|$$

where, as in earlier cases, $|p - p'|$ is the distance between two objects or points in C_i and C_j.

The distance formula may also be written as:

$$Dist(C_1,C_2) = \frac{1}{n_{C_i} + n_{C_j}} \sum_{i=1}^{C_1} \sum_{j=1}^{C_2} Dist(i,j), i \in C_1 \text{ and } j \in C_2$$

where:

- n_{c_1} = Number of element or data-points in C_1
- n_{c_2} = Number of element or data-points in C_2

Depending on the type of linkage, the generated dendrograms would differ. Consider the distribution shown in Fig. 10.2.

The generated dendrograms for the three types of linkages on this dataset are shown in Fig. 10.3.

Though the tree structures created in three cases are different, the clustering effect remains the same.

The project source for the above simulation is available in the book's repository.

I will now discuss a few advantages and disadvantages of agglomerative clustering.

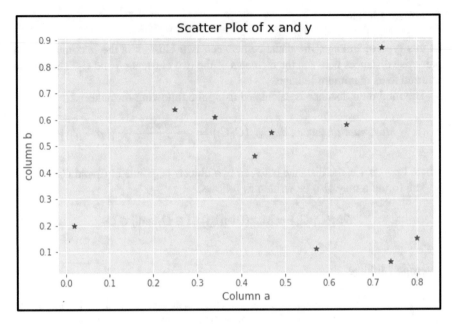

Fig. 10.2 Random data distribution

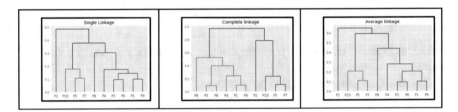

Fig. 10.3 Dendrograms for different linkage types

Advantages and Disadvantages

Here are a few advantages of AGNES:

- You do not have to specify the number of clusters like in the case of *K*-means clustering.
- It is easy to implement and easy to understand.
- The tree hierarchy is many-a-times more informative than the unstructured set of flat clusters returned by *K*-means clustering.

Here are a few disadvantages of AGNES:

- During clustering, the process of cluster merging cannot be reversed. This can become problematic in case of noisy, high-dimensional datasets.
- For big data, the clustering could become computationally expensive.

Finally, I will give you a few use cases where AGNES has given satisfactory results or can be used.

Applications

- Hierarchical clustering is useful in determining the phylogenetic tree of animal evolution.
- You can track viruses by charting phylogenetic trees.
- We can phylogenetically analyze the bacterium of saliva.
- We can use the algorithm for document classification, where similar documents can be organized by analyzing the text similarity.
- We can also use the algorithm in aiding sales and marketing. You can group customers having similar traits and likelihood to your products and services, which you would use for designing your strategies.

Implementation

The sklearn implements agglomerative clustering *sklearn.clustering. AgglomerativeClustering* class. You specify the number of clusters, the *affinity* metric, and the type of *linkage* during instantiation.

```
hc=AgglomerativeClustering(n_clusters=3,
                           affinity="euclidean",
                           linkage="ward")
```

I will illustrate how the agglomerative clustering is done using the above class with a trivial project.

Project

I tried the agglomerative clustering on the mall customer dataset taken from Kaggle. The customer segmentation is done on the basis of their annual income and age.

First, we will do the clustering based on the annual income. This code snippet will form two clusters in the entire dataset to provide us the segmentation of the customers' spending capacity based on their annual income.

```
hc=AgglomerativeClustering(n_clusters=2,
                           affinity="euclidean",
                           linkage="ward")
```

The output is an array that tells us which data point belongs to which cluster—1 or 2. You can plot the clusters using the following code snippet:

```
fig = plt.figure(figsize=(10, 6))
plt.scatter(x[y_hc == 0, 0], x[y_hc == 0, 1],
            s = 100, c = 'red', label = 'Cluster 1')
plt.scatter(x[y_hc == 1, 0], x[y_hc == 1, 1],
            s = 100, c = 'blue', label = 'Cluster 2')
plt.title('Customers -two clusters')
plt.xlabel('Annual Income (k$)')
plt.ylabel('Spending Score (1-100)')
plt.legend()
plt.show()
```

Figure 10.4 shows the output.

As you observe, customers having mid-range (40-55K) of income exhibit average spending capacity. When they do purchases, it is neither too low nor too high.

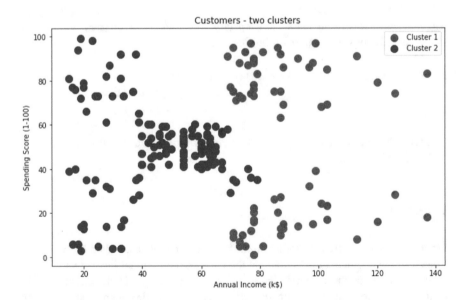

Fig. 10.4 Agglomerative clustering on real-life dataset

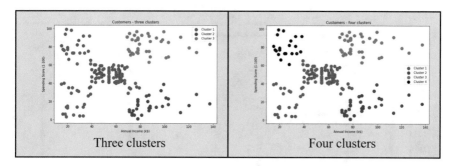

Fig. 10.5 Different cluster sizes with agglomerative clustering

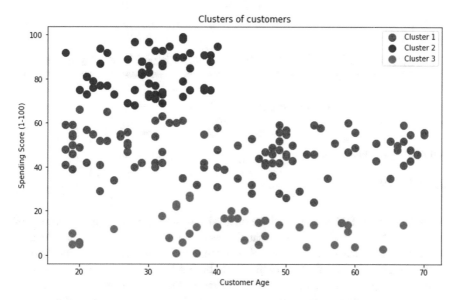

Fig. 10.6 Agglomerative clustering for spending score/customer age

For further investigations, you may try clustering the same dataset in three or four clusters. Figure 10.5 displays the results.

You can now do the analysis by looking at these visualizations to understand the customer's spending capacity depending on the annual income.

You may do a similar analysis based on the customer's age. The 3-clustering output based on the age is shown in Fig. 10.6.

As a curiosity, you can also check the dendrogram. Figure 10.7 shows the output:

The full source for the project is available in the book's repository

Next, I will discuss the divisive clustering.

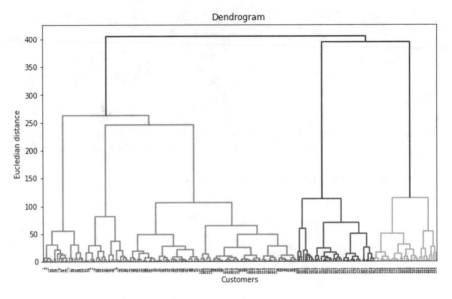

Fig. 10.7   Dendrogram for real-life dataset

Divisive Clustering

In a Nutshell

Also known as DIANA or divisive analysis, this technique is quite similar to agglomerative clustering, except that it uses a bottom-up approach rather than a top-down approach used in AGNES. It begins by putting all data points in one cluster, which becomes the root of the tree to be constructed. The root cluster is then recursively split into smaller clusters until each cluster at the bottom is coherent enough. The clusters at the bottom level are adequately similar to each other or just contain a single element. Figure 10.8 shows the cluster formations in divisive clustering.

The Working

Implementing divisive clustering is not as trivial as AGNES. I will first explain the algorithm and later on discuss the challenges in its implementation. The algorithm has the following simple steps:

1. Put all data points into a single cluster.
2. Partition the cluster into two least similar clusters using some distance measuring technique such as Euclidean distance.

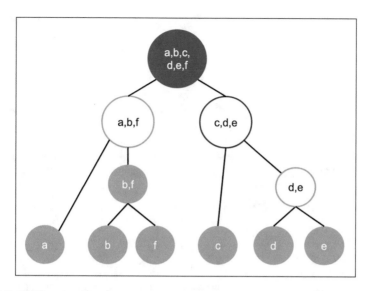

Fig. 10.8 Divisive clustering schematic

Fig. 10.9 Random dataset

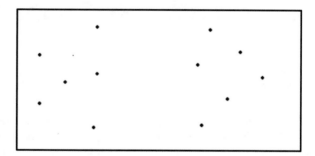

3. Repeat step 2 recursively until there is a single object left in each cluster or at the most the objects which are very similar.

   Diagrammatically, the entire process can be explained as follows:
   Consider the data distribution shown in Fig. 10.9.
   We start by putting all the data points into a single cluster and then split it recursively until we get a set of clusters where each cluster is coherent. Figure 10.10 shows the various stages.
   As you see in the diagram, we have stopped the process after step 4 considering that each innermost cluster is sufficiently coherent. If not, you may continue with further divisions.
   The process looks simple; however, its implementation poses several challenges.

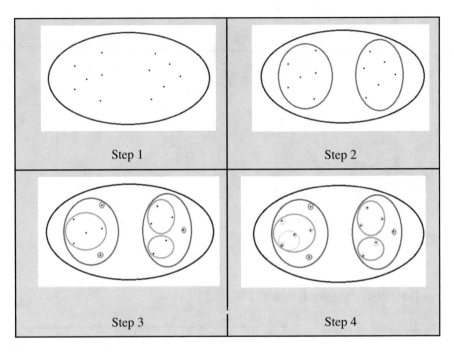

Fig. 10.10 Cluster formation steps

Implementation Challenges

1. The biggest challenge is selecting the appropriate partition while dividing a cluster. Note that there are 2^{n-1}—1 potential ways to partition a set of n objects into two exclusive clusters. When n is large, this can be computationally intensive if you decide to examine all combinations.
2. The algorithm uses heuristics in partitioning that can lead to inaccuracies.
3. The algorithm typically does not backtrack its choices. Once the cluster is partitioned, it does not reconsider an alternative partition and simply proceeds with further splitting.

Due to the above challenges, many times, data scientists prefer agglomerative methods over divisive. The sklearn library does not provide the implementation for divisive clustering.

Summary

In this chapter, you learned two hierarchical-based clustering algorithms—agglomerative and divisive. Agglomerative clustering takes a bottom-up approach, where you start with any arbitrary point in the dataset as a cluster. You then add the nearby points to it by using a proximity metric. You keep on adding more points to this cluster until we visited all points in the dataset. This gives you a hierarchy of all data points in the dataset. You use this hierarchy to decide upon your clusters. Divisive clustering works the other way round—it builds the tree from top to bottom. You start by considering a single cluster of all the data points in the dataset. You divide this cluster using proximity metric. The division continues until you are left with single point clusters at the bottom. Because of the complexity of divisive algorithm, we consider only the agglomerative method of linkage-based clustering.

In the next chapter, you will learn the Gaussian distribution-based clustering.

Chapter 11
Gaussian Mixture Model

A Probabilistic Clustering Model for Datasets with Mixture of Gaussian Blobs

In the previous chapters, you studied centroid-based and connectivity-based clustering. In this chapter, you will study a probabilistic clustering model where you know in advance that the dataset is a mixture of datasets, each having a Gaussian distribution.

In a Nutshell

There are typical situations where you know a dataset would probably contain clusters, each having a Gaussian distribution. Consider, for example, the US population. The distribution of African Americans, Asians, or Native Hawaiian is observed in certain pockets. Assuming that each distribution has an elliptical shape rather than a circular shape, as identified in *K*-means clustering, the Gaussian mixture would be a better approach in identifying these clusters. The *K*-means clustering is a distance-based algorithm that tries to group the closest points together, creating a circular shape for the cluster. The Gaussian mixture model is a distribution-based algorithm where we consider both mean (μ) and the variance (σ) in the distribution.

To understand the algorithm, let us first look at what is Gaussian distribution.

Gaussian Distribution

Figure 11.1 shows the Gaussian distributions for various values of μ and σ.

The algorithm plays with these two parameters to cluster the points. Depending on the values of these parameters, the cluster's shape would differ. You can experiment with distributions using the Matplotlib library. The following code snippet generated the above figure:

P. Sarang, *Thinking Data Science*, The Springer Series in Applied Machine Learning, https://doi.org/10.1007/978-3-031-02363-7_11

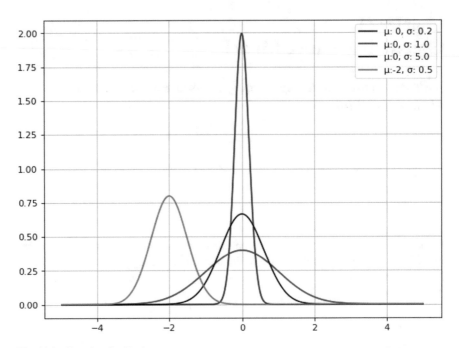

Fig. 11.1 Gaussian distribution

```
plt.plot(x, norm.pdf(x, 0, 0.2), label='μ: 0, σ: 0.2', color = "blue")
plt.plot(x, norm.pdf(x, 0, 1), label='μ:0, σ: 1.0', color = "red")
plt.plot(x, norm.pdf(x, 0, 0.6), label='μ:0, σ: 5.0', color = "black")
plt.plot(x, norm.pdf(x, -2, 0.5), label='μ:-2, σ: 0.5', color = "green")
```

The full source for creating the above plot is available in the book's repository. We now look at probability distribution for determining our clusters.

Probability Distribution

The probability distribution function is stated as:

$$f\left(x|\mu, \sigma^2\right) = \frac{1}{\sqrt{2\pi\sigma^2}} e^{-\frac{(x-\mu)^2}{2\sigma^2}}$$

where μ is the mean and σ^2 represents the variance. This is a distribution function for a one-dimensional space and is valid for a single variable. For a two-dimensional space, we will have a three-dimensional bell curve.

For a multivariate Gaussian distribution, we express the probability density function as:

$$G(X|\mu, \Sigma) = \frac{1}{\sqrt{(2\pi)\mid \Sigma \mid}} \, exp\left(-\frac{1}{2}(X - \mu)^T \Sigma^{-1}(X - \mu)\right)$$

where X represents the input vector, μ is the n-dimensional mean vector, and Σ is the $n \times n$ covariance matrix. We determine the values for μ and Σ by using a statistical technique called expectation-maximization (EM).

Figure 11.2 shows the distributions for various covariance types.

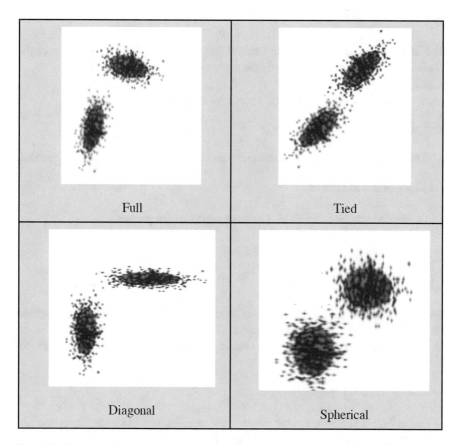

Fig. 11.2 Gaussian distribution for various covariance matrix

Selecting Number of Clusters

You have seen that in *K*-means clustering, we use silhouette score or inertia for getting an estimate of the number of clusters. However, with GMM, the cluster shape may not be spherical. So, we use Bayesian information criterion (BIC) or Akaike information criterion (AIC) to choose the best number of clusters for GMM. I will first show the implementation of GMM as provided by sklearn along with a trivial project for experimentation. Then, I will present a program for estimating the number of clusters using BIC and AIC.

Implementation

The sklearn library provides the implementation of the GMM algorithm in the *GaussianMixture* class. You supply the number of clusters as a parameter.

```
gmm = GaussianMixture(n_components = 3)
gmm.fit(d)
```

After the model is fitted, you discover the clusters by calling its *predict* method.

```
labels = gmm.predict(d)
```

To illustrate the use of GMM, I have created a trivial project.

Project

The project applies the GMM algorithm on two datasets—the famous iris dataset and the random dataset created through code.

The iris dataset is provided in the sklearn library itself. You load it in the project using the following statement:

```
iris = datasets.load_iris()
```

Just in case if you do not remember what it contains, you can print the feature and target names. The output is shown in Fig. 11.3.

Extract the sepal length and width as features for our experimentation.

```
X = iris.data[:, :2]
d = pd.DataFrame(X)
```

```
iris.feature_names
```

```
['sepal length (cm)',
 'sepal width (cm)',
 'petal length (cm)',
 'petal width (cm)']
```

```
iris.target_names
```

```
array(['setosa', 'versicolor', 'virginica'], dtype='<U10')
```

Fig. 11.3 Features and target in iris database

```
array([1, 1, 1, 1, 1, 1, 1, 1, 1, 1, 1, 1, 1, 1, 1, 1, 1, 1, 1, 1, 1, 1,
       1, 1, 1, 1, 1, 1, 1, 1, 1, 1, 1, 1, 1, 1, 1, 1, 1, 1, 1, 0, 1, 1,
       1, 1, 1, 1, 1, 1, 2, 2, 2, 0, 0, 0, 2, 0, 2, 0, 0, 0, 0, 0, 0, 2,
       0, 0, 0, 0, 0, 0, 0, 0, 0, 2, 2, 2, 0, 0, 0, 0, 0, 0, 0, 2, 2, 0,
       0, 0, 0, 0, 0, 0, 0, 0, 0, 0, 0, 0, 2, 0, 2, 0, 2, 2, 0, 2, 2, 2,
       2, 0, 2, 0, 0, 2, 2, 2, 2, 0, 2, 0, 2, 0, 2, 2, 0, 0, 0, 2, 2, 2,
       0, 0, 0, 2, 2, 2, 0, 2, 2, 2, 0, 2, 2, 2, 0, 2, 2, 0])
```

Fig. 11.4 Model predictions

We apply the GMM algorithm on this dataset:

```
gmm = GaussianMixture(n_components = 3)

gmm.fit(d)
```

Now, do the predictions on the original dataset to find out the cluster assignment for each data point.

```
labels = gmm.predict(d)
labels
```

Figure 11.4 shows the predictions made by the model.

As you see, each data point is now associated with some cluster—0, 1, or 2. You can now visualize the clustering using a simple plot routine. Figure 11.5 shows both the original and the clustered datasets.

From the above output, you can understand the dependency of each class of plants on its width and length.

I carried out further experiments on a dataset of 500 data points distributed across three clusters. We create the dataset using the following statement:

```
dataset, clusters = make_blobs(n_samples = 500,
                               n_features = 3,
                               cluster_std = 0.9,
                               random_state = 0)
```

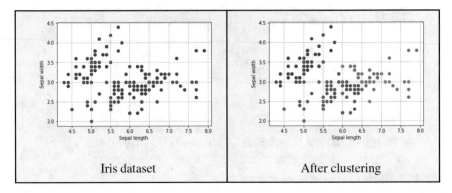

Fig. 11.5   Original dataset and formed clusters

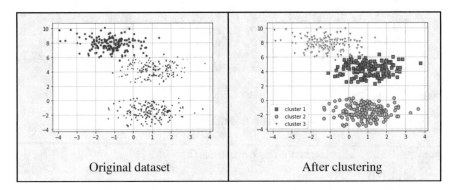

Fig. 11.6   Random dataset before and after clustering

The code for applying GMM and visualizing the cluster datasets remains the same as the earlier one. Figure 11.6 shows the original and the cluster datasets.

The full source of the project is available on the book's download site.

Determining Optimal Number of Clusters

In the *K*-means algorithm, we used silhouette score or inertia to determine the optimal number of clusters. Likewise, in GMM, you use BIC or AIC for estimating the number of clusters. These are the probabilistic model selection techniques that are used for scoring different models. The GMM implementation provides two methods, aic and bic for getting the Akaike and Bayesian information criterion, respectively, for the current model. You use this information for a set of *K* values and then select the best model for final clustering.

Let us first look at Bayesian information criterion (BIC). The GMM algorithm supports four different types of covariance—*full*, *tied*, *diag*, and *spherical*. This is input as one of the parameters in the class instantiation.

```
gmm = GaussianMixture(n_components =k, covariance_type='diag')
```

We will fit GMM on our datasets with these four types of covariance and extract the bic information for different cluster sizes (*K*). The below code snippet shows this.

```
bic_array = []


K=[2,3,4,5]
cv_types = ['spherical', 'tied', 'diag', 'full']
for cov_type in cv_types:
    for k in K:
        gmm = GaussianMixture(n_components =k,
                              covariance_type=cov_type)
        gmm.fit(X)
        bic = gmm.bic(X)
        bic_array.append(bic)
```

After we extract the bic information, we will plot it for different *K* values and covariance. Figure 11.7 shows the output for our dataset.

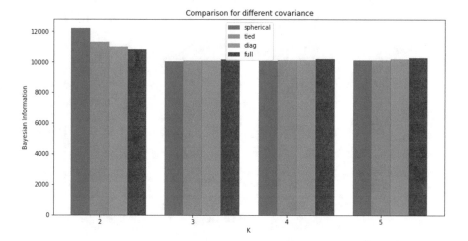

Fig. 11.7 BIC plots

   The height of bars for K equals 3 is lowest among all five groups, showing that the
optimal value for K is 3. The covariance for this dataset does not make much sense.
   We will now plot the aic and bic values for different K. We compute the total of
aic and bic using following code snippet:

```
bic_array = []
aic_array = []


for k in K:
    gmm = GaussianMixture(n_components=k)
    gmm.fit(X)
    bic_array.append(gmm.bic(X))
    aic_array.append(gmm.aic(X))
```

   The plot of aic/bic for different K is shown in Fig. 11.8.
   The elbow in the above chart shows an optimal value for K, which is 3 in our case.

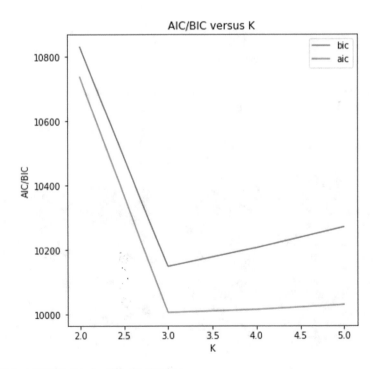

Fig. 11.8 AIC/BIC plots for different K values

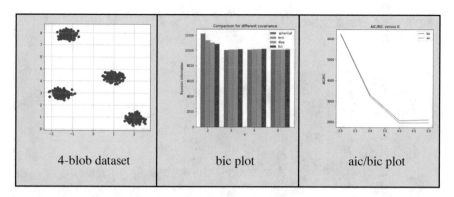

Fig. 11.9 AIC/BIC plots for a four blobs dataset

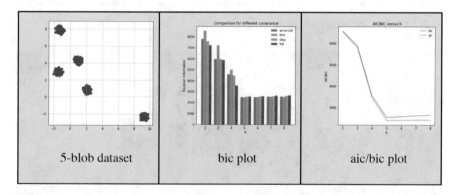

Fig. 11.10 AIC/BIC plots for a five blobs dataset

I carried out a similar experiment on a 4-blob dataset. The results are given in Fig. 11.9.

The aic/bic plot clearly shows the optimal K value equals 4. The bic plots for different covariance types do not give us enough usable information due to the nature of the dataset.

Figure 11.10 presents similar plots for a 5-blob dataset.

As before, the aic/bic plot clearly shows that the optimal value for K equals 5. The bic plot now gives enough variation in bar heights for us to conclude that optimal K equals 5 as we observe that the bic bars for all variance types have the lowest height among the set of all seven observations.

I tried a further experiment on the iris dataset. The results are presented in Fig. 11.11.

Finally, I tried it on the moon dataset created using the *make_moons* function call as follows:

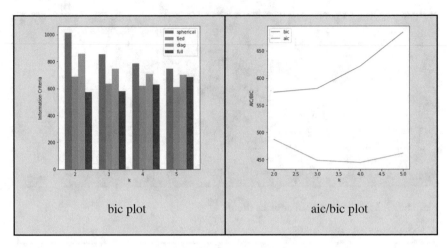

Fig. 11.11 AIC/BIC plots for iris dataset

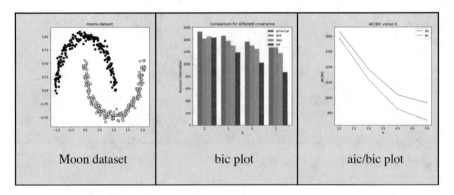

Fig. 11.12 AIC/BIC plots for the moon dataset

```
x2, y2 = make_moons(n_samples=400, noise=0.05,
                    random_state=2021)
```

The results are presented in Fig. 11.12.

You can see that, with iris and moon datasets, the results are not much interpretable for obtaining the optimal value of K. Thus, it is important to understand that this technique of determining optimal K will not work for any kind of dataset. So, a prior knowledge of the data distribution by some other means is essential before applying this technique.

The full source for this project is available in the book's repository for your own experimentation on different types of datasets.

Summary

The Gaussian mixture model's algorithm assumes that your data distribution comprises clusters of Gaussian distributions. To estimate the number of clusters ahead of full clustering, we use aic/bic charts. The charts cannot give us usable information if the distribution is not truly a mixture of Gaussian clusters.

In the next chapter, you will study the density-based clustering.

Chapter 12
Density-Based Clustering

Density-Based Spatial Clustering

In the earlier chapters, you studied flat, hierarchical, and Gaussian mixture model clustering. These algorithms cannot handle outliers or noisy data points. In this chapter, I will introduce you to a clustering technique that is prone to outliers and noise. It is called DBSCAN—density-based spatial clustering of applications with noise. A discussion on another clustering technique called OPTICS that improves upon some shortcomings of DBSCAN will follow this. You will also learn mean shift clustering algorithm.

DBSCAN

In a Nutshell

In this technique, you group the objects having high density of locality into a single cluster. The points with many nearby neighbors form a single group. The points whose nearest neighbors are too far away are treated as outliers.

To illustrate this with an example, consider the case of the US Census data that groups the entire population into several groups (clusters) classified as White Population, Black or African American Population, American Indian and Alaska Native Population, Asian Population and Native Hawaiian, and Other Pacific Islander Population. This can be visualized as seen in Fig. 12.1.

Each cluster will have a high density of similar people. So, the concept in this algorithm is to find objects that have high-dense neighborhoods. These core objects, along with their neighborhoods, form the dense regions to be considered as our clusters.

Importantly, this algorithm is robust to outliers and does not require us to specify the number of clusters beforehand.

© The Author(s), under exclusive license to Springer Nature Switzerland AG 2023
P. Sarang, *Thinking Data Science*, The Springer Series in Applied Machine Learning,
https://doi.org/10.1007/978-3-031-02363-7_12

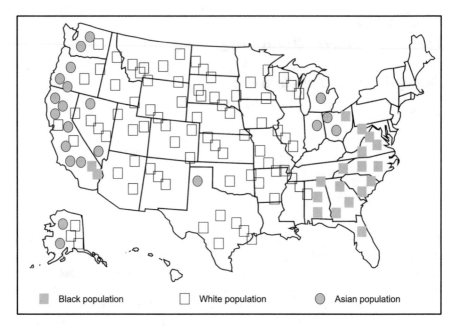

Fig. 12.1 US population clusters

Fig. 12.2 Nonlinear dataset

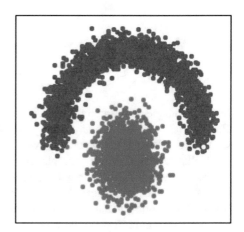

Why DBSCAN?

Consider the data distribution shown in Fig. 12.2.

The cluster separation is nonlinear. The algorithms like the *K*-means and others in that category that you have studied so far cannot appropriately cluster these kinds of datasets. DBSCAN can surely find the nonlinearly separable clusters in such datasets.

To understand the working of the DBSCAN algorithms, you first need to learn a few preliminaries.

Preliminaries

The algorithm works by classifying each data point in the dataset as one of the following three types:

- Core point
- Reachable point
- Outlier

We define ε as the neighborhood radius with respect to a point. If a data point has a *minPts* number of data points within this radius, we classify it as a core point. Figure 12.3 illustrates this.

The point marked in blue color has three other points within a distance of neighborhood radius—ε. This blue color point is called a core point.

We classify a point as a border point if the number of points in its vicinity that is within the ε radius is less than the *minPts*. The point marked in red with a solid circle is such a border point. Note that it has only two neighboring points in the range of ε.

We classify a point as an outlier if there are no other data points within ε radius from it. A point marked in green with an empty circle is such an outlier. Note that in higher dimensions, the circle becomes the hypersphere.

You also need to understand the notion of *reachability* and *connectedness*. Let us consider that p is our core point. Now, if point q lies within distance ε from p, we say that q is *directly reachable* from p. We say that point q is *reachable* from p if the following condition is satisfied. Note that we want to differentiate between *directly reachable* and *reachable* conditions.

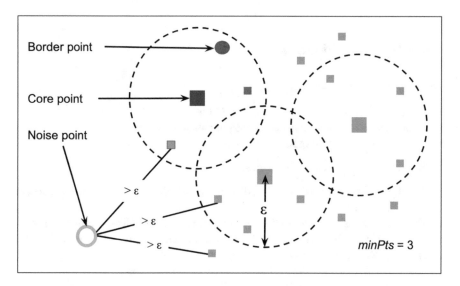

Fig. 12.3 Definitions of core, border, noise points

There exists a path $p_1, \ldots, p_n$ where $p_1 = p$ and p_n equals q such that each p_{i+1} is directly reachable from p_i. This also implies that p and all other points in the chain, except q, are the core points.

You need to understand *connectedness* to formally define the cluster extents. We treat two points p and q as density-connected if there exists another point o such that both p and q are reachable from o. By definition, all points within the cluster are mutually density-connected. If a point is density-reachable from some point in the cluster, it also becomes a part of the same cluster.

With these preliminaries, now let us look at the algorithm steps.

Algorithm Working

These are the steps in the DBSCAN algorithm:

- Decide on the values for *minPts* and ε.
- Arbitrarily pick up a data point in the dataset.
- If there exists a *minPts* number of points in its vicinity, mark it as a core point and begin clustering.
- If the neighboring points are also core points, add all their neighboring points to the same cluster.
- Expand the cluster by recursively following the above procedure.
- Stop the process when all points are visited.

Note that points that are marked as outliers can be part of another cluster. Now, I will list out a few advantages and disadvantages of this algorithm.

Advantages and Disadvantages

These are some of the advantages:

- The algorithm is resistant to noise or outliers.
- It can handle nonlinear clusters of varying shapes and sizes.
- No need to specify the number of clusters.

Here are some of the disadvantages:

- Selecting the proper ε value can be difficult, especially if the data and the scale are not well understood.
- The algorithm is sensitive to the selection of parameter values for *minPts* and ε.
- In case of datasets having large differences in densities, the algorithm may not produce satisfactory clustering.

Implementation

The sklearn library provides the implementation of the DBSCAN algorithm in the *sklearn.cluster.DBSCAN* class. Calling this algorithm is trivial and is done using the following two lines of code:

```
from sklearn.cluster import DBSCAN
db=DBSCAN(eps=3,min_samples=4,metric='euclidean')
```

The three parameters to DBSCAN are self-explanatory. I will now demonstrate its use with a trivial project.

Project

I will use the mall customer segmentation dataset that you used previously in agglomerative clustering. Just to recap, this is quick info on the loaded dataset:

```
Int64Index: 200 entries, 1 to 200
Data columns (total 4 columns):
 #   Column      Non-Null Count   Dtype
---  ------      --------------   -----
 0   Genre       200 non-null     object
 1   Age         200 non-null     int64
 2   Income      200 non-null     int64
 3   SpendScore  200 non-null     int64
dtypes: int64(3), object(1)
```

We will use age and income columns for DBSCAN clustering.

```
X=dataset.iloc[:,[2,3]].values
```

Next, we will apply DBSCAN on this *X* dataset.

```
from sklearn.cluster import DBSCAN
db=DBSCAN(eps=3,min_samples=4,metric='euclidean')
model=db.fit(X)
```

```
array([-1, -1, -1, -1, -1, -1, -1, -1, -1, -1, -1, -1, -1, -1, -1, -1, -1,
       -1, -1, -1, -1, -1, -1, -1, -1, -1, -1, -1, -1, -1, -1, -1, -1,
       -1, -1, -1, -1, -1, -1, -1, -1, -1, -1, -1, -1, -1, -1, -1, -1,
       -1, -1, -1, -1, -1,  0,  0,  0,  0, -1, -1,  0, -1,  0, -1,  0,  0,
       -1,  0, -1, -1,  0, -1,  1,  1,  1,  1,  1,  1,  1,  1,  1,  1,  1,
        1,  1,  1, -1,  2,  1,  2,  2,  2,  2,  2,  2,  2,  2,  2,  2,  2,
        2,  2,  2,  2,  2,  2,  2,  2,  2,  2,  2,  2,  2,  2,  2,  3,  2,
        3,  3, -1,  3, -1, -1,  4, -1, -1, -1,  4,  5,  4, -1,  4,  5, -1,
        5,  4, -1,  4,  5, -1, -1,  6, -1, -1, -1,  7, -1,  6, -1,  6, -1,
        7, -1,  6, -1,  7, -1,  7, -1, -1, -1, -1, -1, -1, -1, -1, -1, -1,
        8, -1,  8, -1,  8, -1,  8, -1, -1, -1, -1, -1, -1, -1, -1, -1, -1,
       -1, -1, -1, -1, -1, -1, -1, -1, -1, -1, -1, -1, -1])
```

Fig. 12.4 Cluster assignment for each data point

When the code execution finishes, you find it assigned each data point to a certain cluster. You can check this by printing the labels assigned to each data point with the following code:

```
label = model.labels_
label
```

Figure 12.4 shows the output array where each element indicates the cluster to which the corresponding data point is assigned.

As you see, for each data point in our dataset of 200, a certain cluster number is assigned. The -1 shows a noisy sample. Notice that the algorithm has automatically decided on the number of clusters. For a large dataset, it will be difficult to examine output similar to above to find out how many clusters it created. You can get this value easily by calling the following statement:

```
number_of_clusters=len(set(label))- (
                              1 if -1 in label else 0)
print('Number of clusters:', number_of_clusters)
```

In the above case this is the output:

```
Number of clusters: 9
```

You can also get the indices of all detected core samples by checking its attribute:

```
db.core_sample_indices_
```

Figure 12.5 shows the output of my run.

As you see the samples at indices 58, 59, 62, and so on are the detected core points.

```
array([ 58,  59,  62,  64,  66,  67,  69,  72,  74,  75,  76,  77,  78,
        80,  81,  83,  85,  87,  91,  92,  93,  94,  95,  96,  97,  98,
        99, 100, 101, 103, 104, 105, 106, 107, 108, 109, 110, 111, 112,
       113, 114, 115, 116, 119, 129, 131, 136, 137, 151, 153, 155, 170])
```

Fig. 12.5 Indices of detected core samples

Fig. 12.6 Number of
features used while
clustering

```
db.n_features_in_
```

2

You may also check upon the number of features used while clustering by checking the value of the attribute *n_features_in_*, which in our case is two, as seen in Fig. 12.6.

Now, we will do the prediction on the same dataset to help us create a visualization plot.

```
y_means = db.fit_predict(X)
```

This statement will associate each data point with a cluster number. We will use this information while plotting each cluster in a different color, as shown in the code snippet here:

```
plt.scatter(X[y_means == 0,0], X[y_means == 0, 1],
                s = 50, c = 'pink')
plt.scatter(X[y_means == 1, 0], X[y_means == 1, 1],
                s =50, c = 'yellow')
plt.scatter(X[y_means == 2, 0], X[y_means == 2, 1],
                s =50, c = 'cyan')
```

A simple scatter plot code will produce the output as seen in Fig. 12.7.

The entire project source is available in the book's repository.

As you have seen, DBSCAN has clustered our dataset without we telling it in advance how many clusters it should create. The algorithm has decided on this number on its own. At the end, we could get this number from the class instance and also visualize the created clusters.

Next, I will describe another important algorithm in this category, and that is called OPTICS.

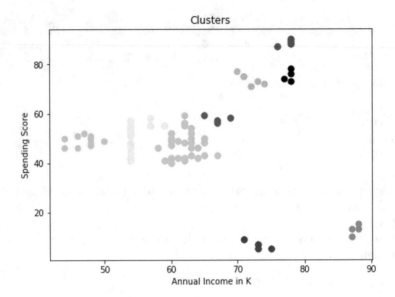

Fig. 12.7 Clustering for spending scores/annual income

OPTICS

Clustering Algorithm to Improve upon DBSCAN

In a Nutshell

A density-based clustering algorithm that addresses one of DBSCAN's major weaknesses. The DBSCAN cannot create meaningful clusters for datasets having varying densities. The OPTICS algorithm addresses this issue and works with datasets having varying densities.

To understand the working of this algorithm, you need to understand two new concepts—*core distance* and *reachability distance*.

Let us first look at the definition of these two additional terms.

Core Distance

In DBSCAN, we defined a core point as a point if at least *MinPts* points are found within its ε-neighborhood. The core distance is now defined as the minimum value of radius required to classify a point as a core point.

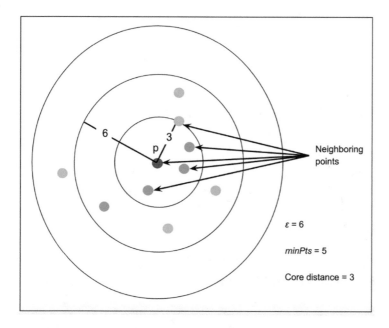

Fig. 12.8 Core point illustration

Figure 12.8 illustrates this.

The point p has five neighbors including itself and lies within the core distance of three units. Thus, p qualifies as the core point.

Reachability Distance

The reachability distance between points p and q is defined as the maximum of the core distance of p and the distance between p and q, with the condition that q be a core point too. Figure 12.9 illustrates the reachability distance, where q is reachable from p with a reachability distance of 7.

This clustering technique differs from others in the sense that it does not explicitly cluster the data; rather it produces the visualization of reachability distances that can help in forming clusters. Figure 12.10 shows such a visualization.

Examining the reachability plot, one can easily see that the dataset will contain three clusters. The clusters show up as valleys in the reachability plot. A deep value shows a dense cluster. I will now discuss the sklearn implementation of this algorithm.

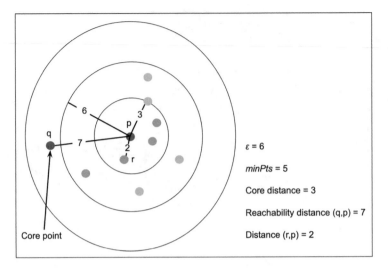

Fig. 12.9 Reachability distance illustration

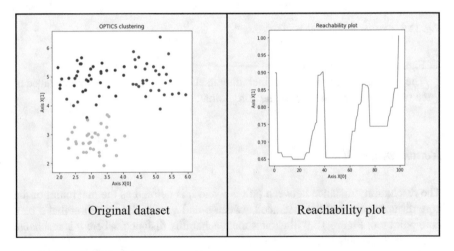

Fig. 12.10 Reachability plot

Implementation

The sklearn library provides the implementation of OPTICS clustering in the *sklearn.cluster.OPTICS* class. Using this class is as trivial as you have seen in the earlier clustering classes provided in sklearn. You just fit your dataset on the OPTICS instance with a single program statement shown here:

```
db = OPTICS(max_eps=2.0, min_samples=20,
            cluster_method='xi,
            metric='minkowski').fit(X)
```

After the algorithm generates clusters in your dataset, you can visualize it by simple plotting routines. I will now show you the project code to show how to do OPTICS clustering and for generating above-shown reachability plots.

Project

For this project, I will use a randomly generated dataset like what we did in some of our earlier projects. First, we will define a few configuration parameters required for the OPTICS algorithm and some values for creating the dataset. This is declared as follows:

```
num_samples_total = 100
cluster_centers = [[3,3], [3,5], [5,5]]
num_classes = len(cluster_centers)
epsilon = 2.0
min_samples = 22
cluster_method = 'xi'
metric = 'minkowski'
```

That is to say, we will create 100 data points with a distribution in three clusters. We specify the cluster centers in the configuration variable. We also declare epsilon, *min_samples*, *cluster_method*, and metric for the use of OPTICS algorithm.

We generate the dataset by calling *make_blobs*.

```
X, y = make_blobs(n_samples = num_samples_total,
                  centers = cluster_centers,
                  n_features = num_classes,
                  center_box=(0, 1), cluster_std = 0.5)
```

Figure 12.11 shows the generated data.
We now apply the OPTICS clustering on the dataset using following statement:

```
db = OPTICS(max_eps=epsilon, min_samples=min_samples,
            cluster_method=cluster_method,
            metric=metric).fit(X)
```

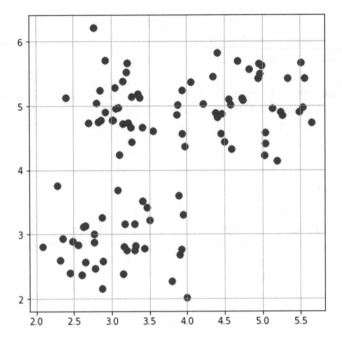

Fig. 12.11 Random dataset

After the clusters are created, we can get its count by examining its attributes as follows:

```
labels = db.labels_
number_of_clusters = len(np.unique(labels) )
number_of_noise_points = np.sum(
                                np.array(labels) == -1,
                                axis=0)

print('Estimated number of clusters: %d' %
      number_of_clusters)
print('Estimated number of noise points: %d' %
      number_of_noise_points)
```

This is the output:

```
Estimated number of clusters: 4
Estimated number of noise points: 12
```

```
[-1  1  0  1 -1  1  1 -1  1 -1  0  1  1  1  1 -1  0  1  0  1 -1  0  1  0
  0  1  0  0  1  1  0  0  1  0  0 -1 -1  1 -1  0 -1  1 -1  0 -1 -1 -1  1
  1  0  0 -1 -1 -1  1  1  0  0 -1 -1 -1  0  0  0 -1 -1  0  0  1 -1  1 -1
 -1  0 -1 -1  0 -1  0  0 -1  1  0  1  1  1  0  1 -1 -1 -1  0 -1  1 -1  0
  0 -1 -1  1]
```

Fig. 12.12 The label assignments for all data points

As expected, it clustered the dataset into three parts. Note that changing those configuration parameters would generate different results. Next, you can do the prediction on the entire dataset for assigning labels to each data point.

```
P = db.fit_predict(X)
```

You can check the label assignment by using the *labels_* attribute:

```
labels = db.labels_
print(labels)
```

Figure 12.12 shows the output where each element displays the label assigned to the corresponding data point.

Each data point is assigned a cluster number to which it belongs. We have three clusters denoted by values -1, 0, and 1.

We can generate a scatter plot of the entire dataset using the following code:

```
# Generate scatter plot for training data
colors = list(map(lambda x: '#ff0000'
                  if x == -1
                  else '#00ff00'
                  if x == 1
                  else '#0000ff', P))
plt.scatter(X[:,0], X[:,1], c=colors, marker="o",
            picker=True)
plt.title('OPTICS clustering')
plt.xlabel('X[0]')
plt.ylabel('X[1]')
plt.show()
```

Figure 12.13 shows the created clusters.

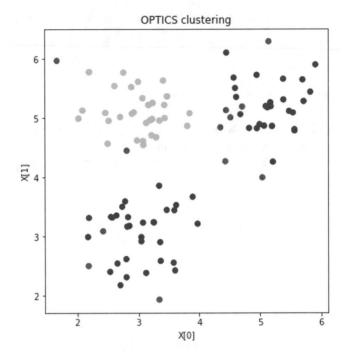

Fig. 12.13 Clusters created by OPTICS

The full source of this project is available in the book's repository.
I will now discuss the next clustering algorithm—mean shift.

Mean Shift Clustering

In a Nutshell

The mean shift clustering algorithm discovers the clusters in a smooth density of data
points. The algorithm begins by defining a circular window centered at some point
C and having the radius r, which is the kernel. The algorithm iteratively shifts this
kernel to a higher density region until convergence.

Algorithm Working

We base the algorithm on the concept of kernel density estimation (KDE). Figure 12.14 shows a typical data distribution and the KDE contour plot.
   So, the problem essentially boils down to discovering blobs by using KDE
functions. The algorithm continually shifts the sliding window closer to a nearby
peak to find the high-density regions. The process is called hill climbing.

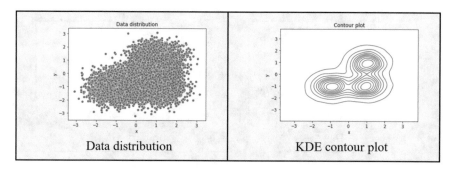

Fig. 12.14 Data distribution and KDE contours

The source code for generating the above image is available in the book's repository for your further experimentation.

These are the steps in the algorithm:

1. Define a sliding window (kernel bandwidth) around a data point.
2. Compute the mean of all points in the window.
3. Move the window centroid to the mean.
4. Repeat steps 2 and 3 until convergence.
5. Delete overlapping windows.
6. A window containing the most points is preserved.

I will now show you the effect of selecting an appropriate bandwidth.

Bandwidth Selection

It is important that you select the right bandwidth, i.e., the radius of the sliding window. Figure 12.15 shows the effect of different bandwidths on our sample dataset.

As you can see, the resulting clusters look different depending on the bandwidth selection. If the bandwidth is too small, each data point will have its own cluster. For a large bandwidth, all data points will belong to a single cluster.

I have added the source code for generating the above image in the book's repository. You may like to experiment on different datasets and bandwidth values for your further understanding of the bandwidth effect on contouring.

The sklearn library provides the estimation function for the bandwidth. Just add the following code snippet in your project to get an estimate for the bandwidth.

```
from sklearn.cluster import estimate_bandwidth

bandwidth = estimate_bandwidth(X, quantile=0.2,
                               n_samples=500)
```

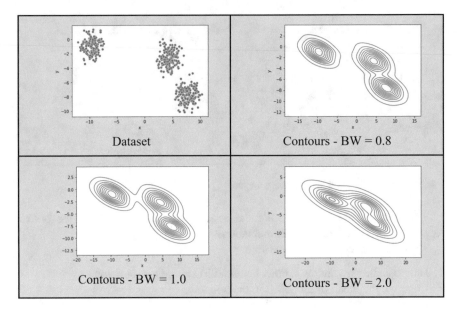

Fig. 12.15 Contours for different values of BW

Obviously, you need to set the appropriate parameter values in the call to *estimate_bandwidth* function based on your dataset.

I will now give you the pros and cons of this algorithm.

Strengths

- No prior assumption on the cluster shape like spherical, elliptical, etc.
- Discovers variable number of clusters—does not require you to define the number of clusters beforehand.
- Robust to outliers.
- Requires only a single parameter—bandwidth.
- Capable of handling arbitrary feature spaces.

Weaknesses

- Bandwidth selection is non-trivial.
- An inappropriate selection of bandwidth can result in merged or shallow regions in the formation of clusters.
- Often some experimentation on bandwidth is required.
- Computationally expensive.
- Does not scale well for high-dimensional feature spaces.

Fig. 12.16 Segmentation by the mean shift algorithm

Applications

One of the primary areas where this algorithm is mainly used is image processing and computer vision. Figure 12.16 is an illustration of image segmentation using the mean shift clustering algorithm.

If you want to try out the image segmentation on your own images, I have made the source code of the project available in the book's repository.

I will now show you the implementation of this algorithm as provided in sklearn.

Implementation

The sklearn library implements mean shift clustering algorithm in the *sklearn.cluster.MeanShift* class. The use of this class is trivial as can be seen in the following code snippet:

```
meanshift = MeanShift(bandwidth=2)
meanshift.fit(X)
```

You just need to specify the appropriate bandwidth. You can get the bandwidth estimation by calling the built-in function.

```
from sklearn.cluster import estimate_bandwidth


bandwidth = estimate_bandwidth(X, quantile=0.2,
                               n_samples=500)
```

After the algorithm creates clusters, you can get statistics using its attributes and visualize the clusters by using any plotting library. I will illustrate these techniques through a trivial project.

```
dataset.info()
```

```
<class 'pandas.core.frame.DataFrame'>
Int64Index: 200 entries, 1 to 200
Data columns (total 4 columns):
 #   Column                  Non-Null Count  Dtype
---  ------                  --------------  -----
 0   Genre                   200 non-null    object
 1   Age                     200 non-null    int64
 2   Annual Income (k$)      200 non-null    int64
 3   Spending Score (1-100)  200 non-null    int64
dtypes: int64(3), object(1)
memory usage: 7.8+ KB
```

Fig. 12.17 Dataset information

Project

For demonstration of this algorithm, I will apply it on two datasets—mall customer dataset obtained from Kaggle and a dataset comprising random data points.

After loading the dataset in memory, you can print its info to understand what it contains. Figure 12.17 is the summary info on the dataset.

The Age and Annual Income are our features and Spending Score is the target. We will do the clustering based on the annual income. We create our training dataset:

```
X = dataset.iloc[:, [2, 3]].values
```

We estimate the bandwidth:

```
ms = MeanShift(bandwidth=bandwidth)
ms.fit(X)
```

We do mean shift clustering:

```
ms = MeanShift(bandwidth=bandwidth)
ms.fit(X)
```

We do the prediction on the original dataset to observe the association of each data point with the cluster number.

```
P = ms.predict(X)
```

Figure 12.18 shows the predictions made by the model.

You see that each data point is associated with a particular cluster number - 0, 1, or 2. You can also get a count of number of clusters by using this code:

P

```
array([0, 0, 0, 0, 0, 0, 0, 0, 0, 0, 0, 0, 0, 0, 0, 0, 0, 0, 0, 0, 0, 0,
       0, 0, 0, 0, 0, 0, 0, 0, 0, 0, 0, 1, 0, 0, 0, 0, 0, 0, 0, 1, 0, 0,
       0, 0, 0, 0, 0, 0, 0, 0, 0, 0, 0, 0, 0, 0, 0, 0, 0, 0, 0, 0, 0, 0,
       0, 0, 0, 0, 0, 0, 0, 0, 0, 0, 0, 0, 0, 0, 0, 0, 0, 0, 0, 0, 0, 0,
       0, 0, 0, 0, 0, 0, 0, 0, 0, 0, 0, 0, 0, 0, 0, 0, 0, 0, 0, 0, 0, 0,
       0, 0, 0, 0, 0, 0, 0, 0, 0, 0, 0, 0, 0, 1, 2, 1, 2, 1, 2, 1, 2, 1,
       2, 1, 2, 1, 2, 1, 2, 1, 2, 1, 2, 1, 2, 1, 2, 1, 2, 1, 2, 1, 2, 1,
       2, 1, 2, 1, 2, 1, 2, 1, 2, 1, 2, 1, 2, 1, 2, 1, 2, 1, 2, 1, 2, 1,
       2, 1, 2, 1, 2, 1, 2, 1, 2, 1, 2, 1, 2, 1, 2, 1, 2, 1, 2, 1, 2, 1,
       2, 1])
```

Fig. 12.18 Model predictions

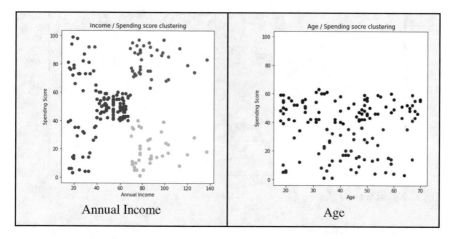

Fig. 12.19 Mean shift clustering on real data

```
number_of_clusters = len(ms.cluster_centers_)
print('Number of clusters: ', number_of_clusters)
```

Finally, you can observe the cluster visualization with a simple plotting routine.

By selecting the Age as a feature in place of Annual Income, you will see the spending capacity based on the customer age. Figure 12.19 shows clustering based on annual income and age.

I will now show you the clustering done on a random dataset. A three-dimensional dataset is created using *make_blobs* as follows:

```
clusters = [[1,1,1],[5,5,5],[3,10,10]]
X, _ = make_blobs(n_samples = 150,
                  centers = clusters,
                  cluster_std = 0.60)
```

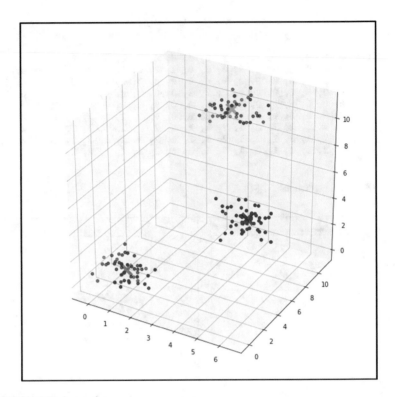

Fig. 12.20 3-D cluster plot

Again, you will get an estimate of bandwidth before applying mean shift. After the dataset is clustered, you can observe the clustering. This entire code is like what we did in the earlier case. Figure 12.20 shows the visualization produced in my test run.

The full source code for this project is available in the book's repository.

Summary

In this chapter, you studied three important clustering algorithms, DBSCAN, OPTICS and Mean Shift that work on datasets having nonlinear density curves. These algorithms are robust to outliers and do not require estimating the number of clusters ahead of clustering. The OPTICS algorithm improves upon the DBSCAN by adding the reachability concept. Thus, it can cluster datasets having varying densities. In the case of the mean shift algorithm, you used the statistical estimation techniques to get an estimate on the number of clusters.

In the next chapter, you will study the BIRCH algorithm.

Chapter 13
BIRCH

Divide and Conquer

In this chapter, you will learn the use of another important clustering algorithm that is called BIRCH—balanced iterative reducing and clustering using hierarchies.

In a Nutshell

BIRCH is an unsupervised data mining algorithm suitable for huge datasets. It does hierarchical clustering by splitting a large dataset into smaller ones, keeping as much information as possible from the original. We then cluster the individual ones using other known clustering techniques. A major limitation of BIRCH is that it works only on metric attributes—attributes whose values we can only represent in Euclidean space. It does not work on categorical values.

Why BIRCH?

The machine learning algorithms previous to the invention of BIRCH lacked the ability to work on vast databases that would not fit in primary memory. The additional input/output operations added the complexity and the cost in achieving high-quality clustering. The earlier algorithms considered every data point equally important while making clustering decisions. BIRCH changed this picture substantially. It makes each clustering decision without scanning all data points. It considers the fact that the data distribution is not usually uniform and all data points are not equally important while clustering. It uses the full memory and does the clustering incrementally that eliminates the need for loading the entire dataset into the memory. BIRCH requires only a single scan of the dataset and does an incremental and dynamic clustering of the incoming data. It can handle noise effectively.

P. Sarang, *Thinking Data Science*, The Springer Series in Applied Machine Learning, https://doi.org/10.1007/978-3-031-02363-7_13

To understand the BIRCH algorithm, you need to understand two terms—CF (clustering feature) and CF tree.

Clustering Feature

BIRCH first summarizes the entire dataset into smaller, dense regions. We call these clustering features. A clustering feature in an ordered triple expressed as:

$$\text{Ordered triple} - (N, LS, SS),$$

where:

N is the total number of data points in a cluster
LS is the linear sum of these data points
SS is the squared sum of data points

A CF entry may contain other CF entries.

CF Tree

The CF tree represents the entire dataset, where each leaf node is a subcluster. Each entry is a pointer to a child node and a CF entry that is the sum of all CF entries in its child nodes. Each leaf node can contain only a maximum number of entries called threshold. Figure 13.1 shows a visualization of the CF tree.

Let me explain the structure of the tree shown in Fig. 13.1. The root node and each of the leaf nodes contain at most B entries, where B is the branching factor. B takes the form [CF_i, child$_i$] where CF_i represents the ith subcluster and child$_i$ is a pointer to its ith child node. A leaf node contains at most L entries. Each entry takes the form [CF_i]. It also contains next and previous pointers to chain all leaf nodes together. The size of the tree depends on the parameter T, the threshold. Note that the entire tree is a compact representation of the dataset, with each entry in a leaf node being a subcluster.

Having understood the two terms and the tree structure, now let us look at the algorithm itself.

BIRCH Algorithm

The algorithm takes two inputs—a set of N data points, represented as real-valued vectors (no categorical attributes), and a desired number of clusters K. It operates in four phases:

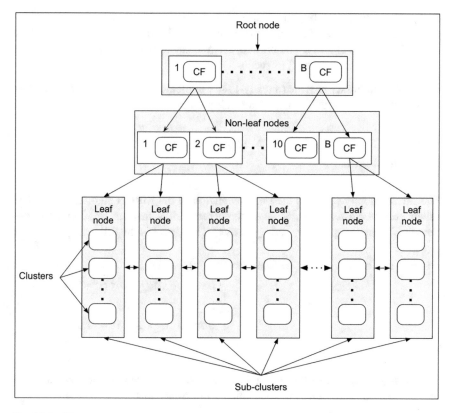

Fig. 13.1 CF tree

1. Build the CF tree.

    (a) Create clustering features with the following definitions:

    $$CF = \left(N, \overrightarrow{LS}, SS\right)$$

    $$\overrightarrow{LS} = \sum_{i=1}^{N} \overrightarrow{X}_i$$

    $$SS = \sum_{i=1}^{N} \left(\overrightarrow{X}_i\right)^2$$

    (b) Organize CFs in a height-balanced tree with branching factor B and threshold T as parameters.

2. This is an optional step, in which you will rebuild the tree into a smaller CF tree, by removing outliers and grouping crowded subclusters into larger ones.

3. Apply existing clustering algorithm to cluster all leaf entries.
4. This step 4 too is optional. This is used for handling minor and localized inaccuracies in clusters. The clusters are re-created by using the centroids of the clusters produced in step 3 as seeds.

The sklearn library provides a ready-to-use implementation of BIRCH. I will now show how to use it with the help of a small project.

Implementation

The sklearn library provides the implementation of the BIRCH algorithm in a class called *sklearn.cluster.Birch*. It takes three parameters that are important to us—*threshold*, *branching_factor*, and *n_clusters*. The class has the following prototype:

```
class sklearn.cluster.Birch(*, threshold=0.5,
                            branching_factor=50,
                            n_clusters=3,
                            compute_labels=True,
                            copy=True)
```

You create the clusters by calling its *fit* method and do the inference by calling its *predict* method. The following code snippet illustrates this:

```
model = Birch()
model.fit(dataset)
```

After the clusters are formed, you can visualize them using a simple plotting routine.
I will now describe its implementation through a trivial project.

Project

For the demonstration, we will create a blobs dataset comprising 5000 points distributed among three features:

```
X, Y = make_blobs(n_samples = 5000,
                  n_features = 3,
                  cluster_std = 5,
                  random_state = 128)
```

We now fit the BIRCH model on the dataset by using the default parameter values.

```
model = Birch().fit(X)
```

We can observe the association of each data point with the cluster by predicting on the original dataset.

```
pred = model.predict(X)
pred
```

This is the output:

```
array([0, 0, 1, ..., 0, 0, 2])
```

Observing this array, we know to what cluster each data point is added.

Figure 13.2 shows the original and clustered distribution.

You may now add the *n_clusters* parameter to the algorithm to see its effect. Figure 13.3 shows three different formations by assigning different values to this parameter.

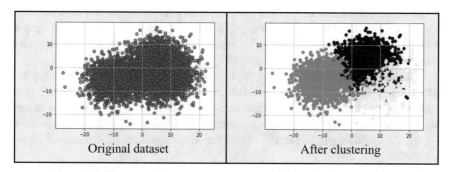

Fig. 13.2 BIRCH clustering

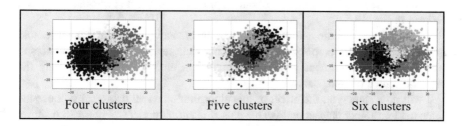

Fig. 13.3 BIRCH clustering for varying number of clusters

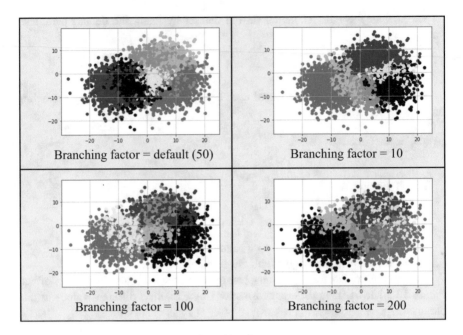

Fig. 13.4 BIRCH clustering for varying branching factor

By default, the value assigned to this parameter is three. When you set a different value for this parameter, the agglomerative clustering model is fitted on the dataset with its *n_clusters* parameter set to be equal to the specified value. You may also provide your own estimator as a parameter.

Now, let us look at the effect of the branching factor. The default value for this parameter is 50. This value determines the maximum number of CF subclusters in each node. During training, if the number of subclusters exceeds this value at any node, it will be split into two independent branches, and the tree is reformed. Figure 13.4 shows the clustering effect for various values of *branching_factor* parameters.

Next, let us look at the effect of *threshold* parameters. This has a default value of 0.5. This value decides on how close any subcluster could be. The closest subcluster should always be less than the threshold. Otherwise, a new subcluster is formed. A very low value for this parameter promotes splitting. The effect of this parameter can be seen in Fig. 13.5.

Finally, to say that the number of clusters, the branching factor, and the threshold are the three parameters that would decide on the clustering. As a data scientist, you would need to experiment with these parameters until it satisfied you with the clusters formed on your dataset.

I will now show you the effect of this algorithm on a real dataset. I will use the same mall customer dataset that we have used in earlier projects. Figure 13.6 shows the clustering done on income/spending data.

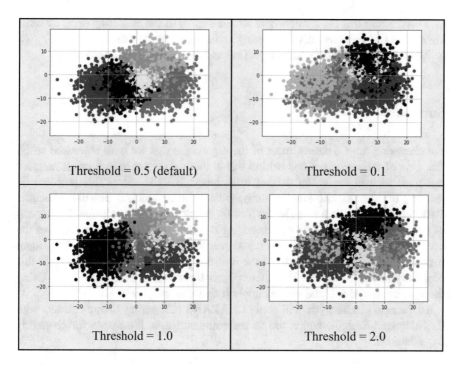

Fig. 13.5 BIRCH clustering for varying threshold values

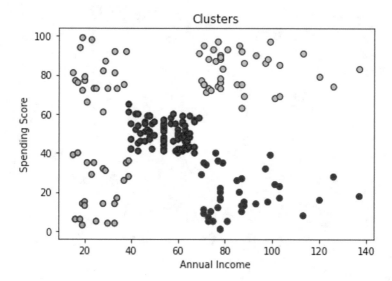

Fig. 13.6 BIRCH clustering on real-life dataset

You observe that the clustering is almost similar to the clustering done by other algorithms like agglomerative clustering and mean shift clustering.

The full source code for this project is available in the book's repository.

Summary

For clustering huge datasets, most of the algorithms that we have discussed so far fail. One of the major reasons behind this is that we do not have enough memory resources to load the entire dataset in memory. BIRCH solves this problem by splitting the dataset into smaller units, creating a hierarchical tree-like structure. Each subcluster that is formed in the process can now be further clustered using any of the earlier algorithms. As the clustering is done incrementally, it eliminates the need for loading the entire dataset into memory. The algorithm has three input parameters, the number of clusters, the branching factor, and the threshold. In this chapter, you studied the effect of these parameters on the cluster formations. Using the algorithm is trivial, and we can apply it easily on huge datasets.

In the next chapter, you will study CLARANS (Clustering LArge datasets with RANdomized Search), which too as the name suggests, is suitable for clustering huge datasets.

Chapter 14
CLARANS

Clustering Large Datasets with Randomized Search

In earlier chapters, you studied many clustering algorithms that work beautifully on small datasets, but fail on huge datasets of sizes million objects and above. In this chapter, you will learn CLARANS, along with its full implementation to cluster datasets of enormous sizes.

In a Nutshell

As the K-medoids algorithm is computationally expensive to execute on large datasets, the CLARA (Clustering for Large Applications) algorithm was introduced as its extension. CLARANS (Clustering for Large Applications with Randomized Search) is a step further that takes care of the cons of K-medoids to handle spatial data. We achieve this by maintaining a balance between the computational cost and the random sampling of data.

CLARA Algorithm

CLARA applies the PAM (partitioning around medoids) algorithm on a small sample of data with a fixed size to generate an optimal number of medoids for the sample. To determine the clustering quality, we measure the average dissimilarity between every object in the entire dataset and its medoid. The sampling and clustering process is repeated a pre-specified number of times to take care of the sampling bias.

The algorithm itself is stated as follows:

Repeat the following procedure for a fixed number of times:

P. Sarang, *Thinking Data Science*, The Springer Series in Applied Machine Learning, https://doi.org/10.1007/978-3-031-02363-7_14

1. Select multiple subsets of fixed size from the original dataset.
2. Apply the PAM algorithm on each subset to determine its medoid.
3. Compute the mean of the dissimilarities of the observations to their closest medoid.
4. A sub-dataset for which the mean is minimal is retained as the final clustering.

The shortcoming of this algorithm is that if the sampled medoids are away from the best medoids, the clustering will not be proper. Raymond Ng and Jiawei Han introduced the CLARANS algorithm in October 2002 that adds randomness in the selection process. I will now describe the CLARANS algorithm.

CLARANS Algorithm

CLARANS works on a trade-off between the cost and effectiveness of using samples.

The CLARANS algorithm requires three input parameters

- *numlocal*—the number of iterations for solving the problem
- *maxneighbor*—the maximum number of neighbors to be examined
- *k*—the number of clusters to be formed

The algorithm itself is stated as follows:

1. Start with the first iteration ($i = 1$). We continue the iterations until $i < numlocal$.
2. Initialize the *mincost* to a large number (infinity) and the *bestnode* to an empty tuple that is used for recording optimal medoids.
3. Randomly select k data points as current medoids. Form the clusters around these medoids. While forming the clusters, consider only the *maxneighbor* of data points. When this limit is crossed, compare the cost of this *current* with the *mincost*. If the *current* cost is less than the *mincost*, set the new value for *mincost*, which is the cost of the *current*. Note that we have set the minimum cost (*mincost*) to infinity at the start. So, the first time, this swap definitely occurs. Also, set the *bestnode* to the *current* node as so far this is the best we have found.
4. Increment i and repeat the entire above process from steps 2 through 4 until i is less than pre-specified *numlocal*.
5. At this point, the *bestnode* tuple contains the data on the best clustering.

From the above discussions, you can make out that taking a high value for *maxneighbor*, the algorithm approaches PAM and it will take longer time to search a local minima. This also results in fewer local minima with improved quality. Thus, the appropriate selection of values for these parameters will result in the trade-off between the quality of clustering and the requirements on the computational resources. Incidentally, CLARANS is a main-memory clustering technique, so conserving the resources becomes an important consideration in parameter selections.

I will now list out some of the major advantages of this algorithm.

Advantages

Here are a few advantages:

- CLARANS is more efficient than PAM and CLARA. While CLARA checks every neighboring node, CLARANS does it only for *maxneighor*, resulting in smaller number of searches as compared to CLARA.
- Ideally suited for spatial datasets of large sizes.
- Compared to other methods (hierarchical, partitioning, density-based, and grid-based), this method scales better on large datasets and even with increasing dimensionality. Note that it does not use any auxiliary structures such as trees or grids and is totally based on randomized search.
- As it is based on local search techniques, unlike others which use Euclidean distance function, it can use any other arbitrary distance measure function.
- While other techniques deal with point objects, CLARANS supports polygonal objects.
- In general, CLARANS gives higher clustering quality than CLARA.

Just to mention, there is also a disadvantage. As CLARANS uses main-memory, the algorithm's efficiency is compromised whenever extensive I/O operations are involved. However, the authors claim that clustering 1 million objects would require slightly over 16 Mbytes of primary memory, which is easily available on even personal computers.

The sklearn does not provide the CLARANS implementation. So, I have taken the implementation provided in the PyClustering library to demonstrate its use. To apply CLARANS on the dataset, you use the following command:

```
clarans_obj = clarans(wine_data, number_clusters = 2,
                      numlocal = 3, maxneighbor = 5)
```

The first parameter is the reference to your dataset. The *number_clus*ters specifies the desired number of clusters. The *numlocal* parameter specifies the number of local minima, which is the number of iterations for running the algorithm. The *maxbeighbor* specifies the maximum number of neighbors to be tested.

As the *clarans* processing typically takes a long time to complete, the designers have provided a method called *timedcall* that at the end tells you the time required for the execution. You invoke the *clarans* process by passing it as an argument to *timedcall* as shown here:

```
(execution_time, res) = timedcall(clarans_obj.process);
```

After execution completes, you will examine the value of *execution_time* to know the time taken by the algorithm process. The process method starts the clustering process. After it completes the clustering, you will obtain the number of clusters, the medoids, and the best medoid.

I will now show the above syntax through a trivial project applied to a real-life dataset.

Project

For this project, I will use the wine dataset provided in sklearn library itself. This is a small dataset and thus the algorithm does not take too much processing time to cluster it. You load the dataset with the following command:

```
dataset = datasets.load_wine()
```

The dataset comprises 13 features which decide on the wine quality. The features are listed using the following statement:

```
dataset.feature_names
```

This is the output:

```
['alcohol',
 'malic_acid',
 'ash',
 'alcalinity_of_ash',
 'magnesium',
 'total_phenols',
 'flavanoids',
 'nonflavanoid_phenols',
 'proanthocyanins',
 'color_intensity',
 'hue',
 'od280/od315_of_diluted_wines',
 'proline']
```

You can check the target column using the following statement:

```
dataset.target_names
```

This is the output:

```
array(['class_0', 'class_1', 'class_2'], dtype='<U7')
```

After you have understood the dataset design, let us apply the CLARANS algorithm on it for clustering. This is done with the following code snippet:

```
clarans_instance = clarans(wine_data, number_clusters = 2,
                    numlocal = 3, maxneighbor = 5)
(execution_time, res) = timedcall(
                        clarans_instance.process);
```

It took 32 seconds to execute the above code. Now, we can find out the allocation of data points to the formed clusters. You do so by calling *get_clusters* method as follows:

```
clusters = clarans_instance.get_clusters()
print("Data points in cluster 1:\n",clusters[0])
print("Data points in cluster 2:\n",clusters[1])
```

This is the partial output:

```
Data points in cluster 1:
 [0, 1, 2, 3, 5, 6, 7, 8, 9, 10, 11, 12, 13, ...]
Data points in cluster 2:
 [4, 20, 21, 39, 43, 59, 60, 61, 62, 63, 64, ...]
```

You can also get the associated target class for each data point by examining the *target* attribute.

You can also get the visualization of the clusters in different dimensions by using the multidimensional visualizer provided in the library.

```
visualizer = cluster_visualizer_multidim()
visualizer.append_clusters(clusters,wine_data,
                          marker="*",markersize=5)
visualizer.show(pair_filter=[[1,2],[1,3],[1,4]],
              max_row_size=1)
```

Figure 14.1 shows the output of multidimensional visualizer.

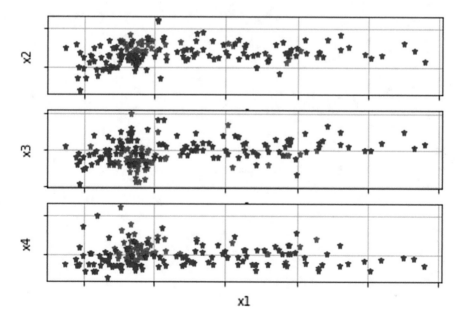

Fig. 14.1 Multidimensional clustering by CLARANS

The output shows the clustering in x1 versus x2, x3, and x4 dimensions. You can easily extend the code to include visualizations in other dimensions.

The full source for this project is available in the book's repository.

Summary

In this chapter, you studied the CLARANS algorithm. CLARANS, which was introduced as an extension to CLARA, is a clustering algorithm for large applications that employ randomized search while handling spatial data. It maintains a balance between computational cost and the random sampling of data. Thus, it is ideally suited for clustering spatial datasets of large sizes. Unlike others, it also supports polygonal objects. It allows you to use any arbitrary distance measuring function. It provides a high clustering quality. The PyClustering library provides its implementation.

In the next chapter, you will learn affinity propagation clustering algorithm.

Chapter 15
Affinity Propagation Clustering

A Gossip-Style Algorithm for Clustering

In a Nutshell

In social networking, when an enormous set of people come together, we split the entire set by forming groups. Each group has a designated leader, and members of each group have a certain affinity toward each other and their leader. The groups are created through peer messaging. That is to say, each member chats with others to decide whether their interests match, whom they like most as the group leader, the availability of the person to accept the responsibilities of a leader, etc. to form the groups. The same idea is picked up while designing the affinity propagation clustering. The algorithm does not require you to determine beforehand the number of clusters to be formed.

Algorithm Working

As an input to the algorithm, we first need to determine the similarity between the data points. We compute the similarity matrix:

Consider a dataset of n points $d_1, \ldots, d_n$.

The similarity matrix is an $n \times n$ matrix where each point $s(i, j)$ represents the similarity between d_i and d_j. The s itself is defined as:

$$s(i, j) = - \| x_i - x_j \|^2$$

where x_i and x_j are the two points.

The matrix diagonal, i.e., $s(i, i)$, decides how likely a particular input is to become an exemplar—a leader for forming the group.

With the above input, the algorithm proceeds by alternating between two message-passing steps, updating two matrices called responsibility and availability

P. Sarang, *Thinking Data Science*, The Springer Series in Applied Machine Learning,
https://doi.org/10.1007/978-3-031-02363-7_15

in each iteration. Each element $r(i, k)$ of the responsibility matrix quantifies how well-suited x_k to be an exemplar for x_i, as compared to other candidate exemplars for x_i. Each element $a(i, k)$ of the availability matrix represents the appropriateness of x_k as an exemplar for x_i.

Initially, both matrices are initialized to zeros. The algorithm then performs the updates on these two matrices iteratively. We run the algorithm for a pre-decided number of iterations.

Responsibility Matrix Updates

First, the responsibility matrix is updated using the following equation:

$$r(i, k) \leftarrow s(i, k) - \underset{k' \neq k}{max} \{a(i, k') + r(i, k')\}$$

where $r(i, k)$ represents the responsibility that point k is the exemplar for point i. How is this decided? The first is the similarity $s(i, k)$ between points i and k. Higher value of s suggests that k would be the i's exemplar. This is not sufficient to determine the responsibility as you also need to consider the suitability of other points as exemplars for i. This is determined by the second term in the above equation. The second term computes the maximum of the sum of availability and similarity of all other samples k'. This max value represents the biggest competitor for k. We subtract this from the first term to compute its relative strength as an exemplar.

Availability Matrix Updates

Each term of the availability matrix is defined as:

$$a(i, k) \leftarrow min \left[0, r(k, k) + \Sigma_{i's.t.i' \notin i, k} r(i', k) \right]$$

The availability is determined as the minimum of zero and the second term of the equation which consists of two parts. The first part $r(k, k)$ determines the responsibility of k to itself, i.e., how important it considers itself to be an exemplar. The second part computes the sum of the responsibilities of all other samples i' to k, where i' is neither i nor k. In simple words, this means a sample will be considered as a potential exemplar if it itself thinks that it is highly important and so do the other samples around it.

In each iteration, we update the scores using the above-defined equations.

Updating Scores

The scores are updated using the following equations:

$$r_{t+1}(i,k) = \lambda \cdot r_t(i,k) + (1 - \lambda) \cdot r_{t+1}(i,k)$$
$$a_{t+1}(i,k) = \lambda \cdot a_t(i,k) + (1 - \lambda) \cdot a_{t+1}(i,k)$$

In every update, the λ of the old value is merged with $+(1 - \lambda)$ of the new value. The λ is a smoothing factor, also called a damping value, ensuring a smooth transition.

After a certain number of iterations, we stop the algorithm and check for the formed clusters.

Few Remarks

From the above discussions, you can see that the algorithm is computationally intensive. The time complexity is $O(n\char`^2 * log\ n)$ and thus should be applied only on small- and medium-sized datasets. The major benefit of this algorithm is that it estimates the number of clusters on its own. It is relatively insensitive to outliers.

Though it gives an excellent performance in most of the cases, it cannot apply to non-convex datasets as it easily draws itself into local optima. If you do not set the proper preferences while clustering, it may cause a set of more or fewer clusters than what you would have expected. We then needed a careful analysis of the accuracy scores to determine the most appropriate number of estimated clusters. I have explained this point in depth later in the chapter's project.

The Hierarchical Affinity Propagation is a variant that deals with the quadratic complexity of this algorithm by splitting the dataset into subsets and then performing two levels of clustering.

I will now discuss its implementation as provided in sklearn library.

Implementation

The sklearn library implements affinity propagation clustering algorithm in *sklearn. cluster.AffinityPropagation* class. Using it is trivial:

```
model = AffinityPropagation().fit(X)
```

After the model clusters your dataset, you get the number of clusters found using this code:

```
center_indices = model.cluster_centers_indices_
number_of_clusters = len(center_indices)
```

The model provides several more attributes to determine the homogeneity of the formed clusters, completeness score, adjusted mutual information, adjusted random index, and even silhouette coefficient. You can use this metrics to determine the clustering quality. I will now describe its use with the help of a trivial project.

Project

I will demonstrate the application of the affinity propagation clustering algorithm on two types of datasets—a random dataset created by coding and a real-life dataset.
The random dataset is created using the usual *make_blobs* function call.

```
centers = [[1,1],[-1,-1],[1,-1]]
X, Y = make_blobs(
    n_samples=300, centers=centers,
    cluster_std=0.5, random_state=0
)
```

Figure 15.1 shows the data distribution in my test run.

Fig. 15.1 Random dataset

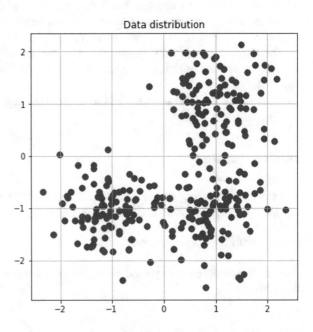

We apply the algorithm using the following statement:

```
model = AffinityPropagation(preference=-50).fit(X)
```

Note that I have set a certain parameter called *preference* in the above call. The value of this parameter plays an important role in cluster creation. I will show you shortly the effect of changing the value of this parameter. With larger values for this parameter, the number of exemplars increases and thus the number of clusters. When you do not specify this parameter in your function call, a median of the input similarities is used as a preference for each point.

After the model performs the clustering, we determine the number of clusters by counting the number of clusters. We check the homogeneity by using the *homogeneity_score* attribute:

```
labels = model.labels_
print("Homogeneity: %0.3f" %
      metrics.homogene-ity_score(Y, labels))
```

In my test run, I got the following value for homogeneity:

```
Homogeneity: 0.872
```

This value surely gives me the confidence that the clustering quality is really good.

You can check the labels array to understand the association of each data point to a particular cluster. Figure 15.2 shows the label values in my test run.

You see that each point is associated to one of the clusters—0, 1, or 2.

You can now plot the clustered dataset with a simple plotting routine. The result of test run is shown in Fig. 15.3.

```
array([0, 1, 2, 0, 0, 2, 1, 1, 2, 0, 1, 0, 0, 1, 0, 1, 1, 1, 1, 0, 0,
       0, 0, 1, 0, 1, 2, 1, 2, 2, 1, 0, 2, 1, 2, 2, 2, 2, 0, 0, 0, 1, 0,
       2, 1, 1, 1, 1, 0, 0, 0, 1, 2, 2, 2, 2, 0, 2, 0, 1, 2, 2, 1, 1, 1,
       2, 1, 2, 0, 2, 1, 0, 0, 1, 1, 2, 2, 0, 1, 2, 0, 1, 2, 2, 0, 0, 1,
       1, 0, 2, 0, 2, 0, 2, 0, 2, 1, 0, 1, 0, 0, 1, 0, 0, 2, 1, 2, 2,
       2, 2, 2, 1, 0, 2, 0, 1, 2, 1, 2, 2, 2, 2, 2, 2, 0, 1, 1, 1, 0, 1,
       1, 2, 0, 2, 0, 1, 2, 0, 1, 1, 2, 0, 1, 0, 2, 1, 2, 2, 2, 0, 0, 0,
       2, 0, 2, 0, 0, 0, 0, 1, 2, 0, 1, 1, 1, 2, 0, 0, 0, 1, 1, 0, 0, 0,
       0, 1, 1, 1, 1, 0, 0, 2, 1, 0, 2, 1, 0, 2, 2, 2, 0, 1, 2, 1, 0, 1,
       0, 1, 0, 0, 1, 2, 2, 2, 2, 0, 1, 1, 0, 2, 2, 2, 0, 1, 0, 0, 1, 0,
       2, 2, 0, 2, 1, 1, 1, 0, 0, 2, 1, 1, 1, 2, 1, 0, 2, 2, 0, 2, 2, 1,
       0, 1, 1, 0, 0, 1, 0, 1, 1, 2, 0, 1, 1, 0, 2, 2, 2, 2, 0, 0, 0,
       0, 0, 1, 1, 2, 0, 0, 2, 2, 2, 1, 2, 1, 2, 2, 1, 1, 1, 2, 1, 0, 1,
       2, 0, 1, 2, 2, 1, 2, 2, 2, 1, 0, 2, 1, 1])
```

Fig. 15.2 Cluster assignments for various data points

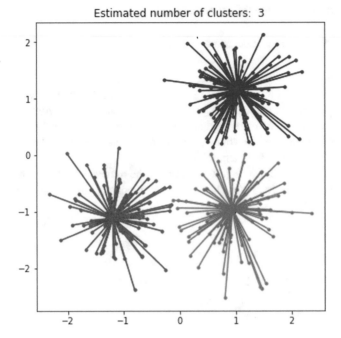

Fig. 15.3 Clusters created by affinity propagation

Next, I will show you the effect of preference value on the clustering. We will simply change the parameter value in the function call as shown in the following statement and plot the clustered data:

```
model = AffinityPropagation(preference=-40).fit(X)
```

The results of different preference values is shown in Fig. 15.4.

As you see, increasing the preference value creates more clusters. This is quite obvious; as I said earlier those larger values of preference would make each point competing to be an exemplar. The interesting case to observe in the above diagram is when preference is set to −2. With such a large value for preference, almost every point gains a good chance of being an exemplar, so 151 points out of 300 have become exemplars. The last and the most important case is when we set the preference to zero. Here, every point has become an exemplar as you observe that the number of clusters formed is 300—the size of our dataset.

One last observation that I would like to present is what happens when you do not set a value for this parameter. As I said earlier, in that case a median of the input similarities would be used as the value for this parameter. The result of using this default value in my test run is shown in Fig. 15.5.

With the default value, the algorithm has now formed 15 clusters in our dataset, which may not be correct as we set the blobs for three clusters. If we had provided a

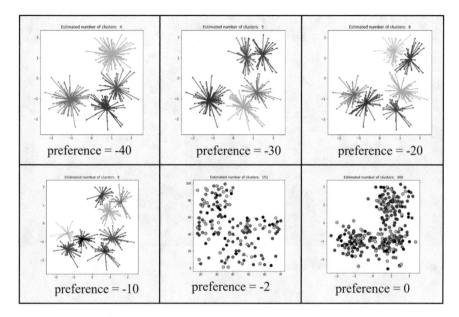

Fig. 15.4 Clustering for varying values of preference

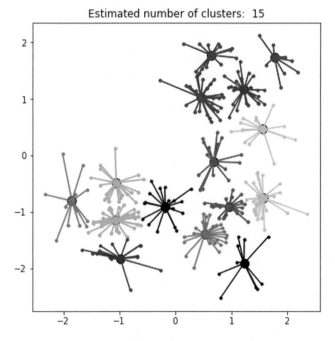

Fig. 15.5 Clustering with default value of preference

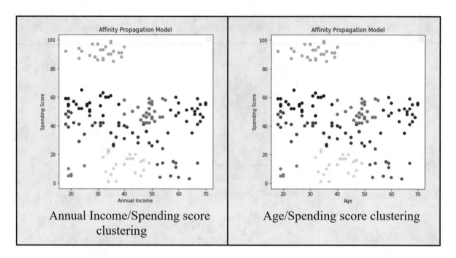

Fig. 15.6 Clustering on real-life data

more separation between the center points while creating the three blobs, the results
would have been closer to this value.

All said, it is very important to set the appropriate value for preference in your
model building. You can attempt different values and then check the homogeneity of
the formed clusters to get your desired quality.

I will now show you the operation of this algorithm on the mall customer dataset
that you have used in your earlier examples. Figure 15.6 gives you the two clustering
plots—one for annual income versus spending score and the other one for age versus
spending score.

The full source code for this project is available in the book's repository.

Summary

The affinity propagation clustering is based on the affinity each data point has to all
other data points in the dataset. Each data point in the dataset tries to be a leader to
form his own group. Only a few succeed and that is how we form the clusters or the
groups in the entire set. The algorithm works by creating two matrices—responsibility and availability. Both matrices are updated over several iterations until we
reach a certain point beyond which further clustering is not possible. The algorithm
is surely computationally expensive, but produces excellent results in many situations, especially when you do not have a knowledge of the number of clusters that a
dataset may have.

In the next chapter, you will study two clustering algorithms called STING and
CLIQUE.

Chapter 16
STING & CLIQUE

Density and Grid Based Clustering

STING: A Grid-Based Clustering Algorithm

In a Nutshell

The clustering algorithms for spatial data that you have studied so far are query dependent, in the sense that you scan the entire dataset for each query. After the query is answered, we cannot reuse the intermediate results of the scanning process for another query. So, these clustering techniques are computationally more complex—$O(n)$. The STING algorithm can answer the queries without re-scanning the entire dataset.

The STING stands for STatistical INformation Grid. I will now show you how this grid is constructed and used for querying.

How Does It Work?

Initially, the entire spatial area is divided into rectangular cells as seen in Fig. 16.1.

As you see in Fig. 16.1, the cells at different layers differ in their resolutions, with each high-level cell containing a summary of the lower level set of cells. We organize the layers in a tree structure, facilitating a quick search on a query. Thus, you can see the computational complexity will be $O(K)$ where K is the number of grid cells at the lowest level and certainly $K << n$, n being the number of data points.

In the beginning, we build the layer hierarchy by starting at the bottom, which contains all the points in our dataset. We use the statistical properties like mean, standard deviation, min, max, and type of distribution like normal, uniform, Gaussian, etc. to group the data points into a cell. Next, compute the cell parameters at the next (higher up from the bottom layer) using these parameters of a lower level cell. Repeat this process until you reach the root layer shown in the diagram. Once the tree

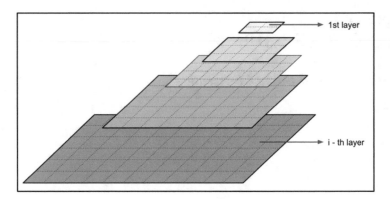

Fig. 16.1 STING layers

hierarchy of cells is built, you use it for querying a new data point. We query using the following algorithmic steps:

- Start at the root layer and proceed downward using the STING index.
- Check the relevance of each cell in this layer to the query by using the statistical information of the cell to build confidence level. Select the cell with the highest confidence level.
- Proceed to the next layer considering the children of the above selected cell.
- Repeat the process recursively until you reach the bottom layer.
- You can now output all the regions at various levels that have satisfied the query.

Advantages and Disadvantages

The advantage of this clustering technique is that the structure is query independent; the entire process can be easily parallelized, and also it is easily possible to provide incremental updates. The complexity too, as said earlier, is low. The disadvantage can be because of its probabilistic nature; you may observe some loss of accuracy in query processing.

Applications

There are several use cases where the grid-based algorithm like STING can be successfully applied. One such query can be finding a region containing a set of houses matching a certain criterion. The condition could be that the region must have a minimum of 100 houses with 80% of it having the price tag about a million dollars. That is, you are interested in finding the regions of high-profile residents. Given a spatial dataset of houses in the entire United States, you can build a layer hierarchy using STING. Then firing such queries would identify all regions in the country that meet the desired criteria.

I will discuss another important algorithm in this category, which is called CLIQUE.

CLIQUE: Density- and Grid-Based Subspace Clustering Algorithm

In a Nutshell

CLIQUE is a subspace clustering algorithm that uses a bottom-up approach to find clusters. It starts with one-dimensional subspaces and keeps merging them to compute high-dimensional subspaces. We consider it both density based and grid based. It provides better clustering in case of high-dimensional datasets.

How Does It Work?

The CLIQUE algorithm combines density- and grid-based clustering. It uses the APRIORI kind of technique to find clusterable subspaces. It starts clustering in a low dimension and then starts connecting the dense subspaces to form clusters in a subspace.

We base it on the following principles:

- We partition each dimension into the same number of equal-length clusters.
- The entire m-dimensional space is partitioned into non-overlapping rectangular units.
- A unit is dense if the number of data points it contains exceeds the present threshold.
- We define a cluster as a maximal set of connected dense units within a subspace.

This is how the algorithm works:

- Start with a one-dimensional subspace.
- Identify the dense subspaces in it.
- We sort these subspaces by coverage, where coverage is the percentage of data points within a subspace with respect to the dataset.
- Keep the subspaces with greatest coverage and prune the rest.
- Find the adjacent dense grid units in each selected subspace using a depth first search.
- Form the cluster by combining these units.

Start the above algorithm with an arbitrary dense unit. Grow it to a maximal region in each dimension. The union of all the regions forms the cluster.

Pros/Cons

Here are the few advantages:

- A simple algorithm.
- Scales linearly with the size of the input.
- Scales well with the increased number of dimensions $O(C^k + mk)$.
- Automatically identifies the subspaces of a high-dimensional data space.
- Does not presume any canonical data distribution.
- Insensitive to the order of records in the input.
- Results are easily interpretable.

The major weakness is that the result quality depends on the choice of the number and width of partitions and grid cells. I will show you the effect of these parameters on the project discussed next.

Implementation

The PyClustering library provides the implementation of the CLIQUE algorithm in the *pyclustering.cluster.clique.clique* class. You just create an instance of clique by passing the data, the number of intervals, and the threshold value for forming a dense subspace:

```
clique_instance = clique(data, intervals, threshold)
```

Then, you create clusters by calling the process method on the created instance.

```
clique_instance.process()
```

After it finishes clustering, discover the clusters by calling get_clusters method:

```
clusters = clique_instance.get_clusters()
```

The library also provides methods to visualize the formed clusters. I will show how to use this algorithm with the help of a trivial project.

Project

Install the PyClustering library and import the required libraries.

```
!pip install pyclustering
from pyclustering.cluster.clique import clique,
                                        clique_visualizer
```

I created a random dataset of 5000 data points of two dimensions. The dataset was created by an online tool. Then I created an instance of clique and formed the clusters in it.

```
# number of cells in grid in each dimension
intervals = 100
threshold = 5    # for deciding on outliers
clique_instance = clique(data, intervals, threshold)
clique_instance.process()
```

The number of formed clusters is displayed by calling *get_clusters*.

```
clusters = clique_instance.get_clusters()
print("Number of clusters:", len(clusters))
```

The output in my run was 121. The grid and cluster plots were created using the built-in visualizer:

```
noise = clique_instance.get_noise()
cells = clique_instance.get_cells()
clique_visualizer.show_grid(cells, data)
```

Figure 16.2 shows the visualizations in my test run.
I experimented with different values for intervals and threshold. Figure 16.3 shows the results of my experimentations.

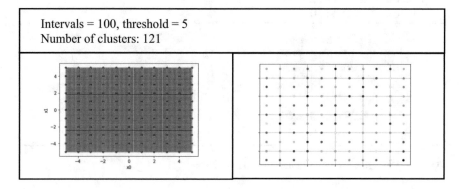

Intervals = 100, threshold = 5
Number of clusters: 121

Fig. 16.2 Grid and cluster visualizations

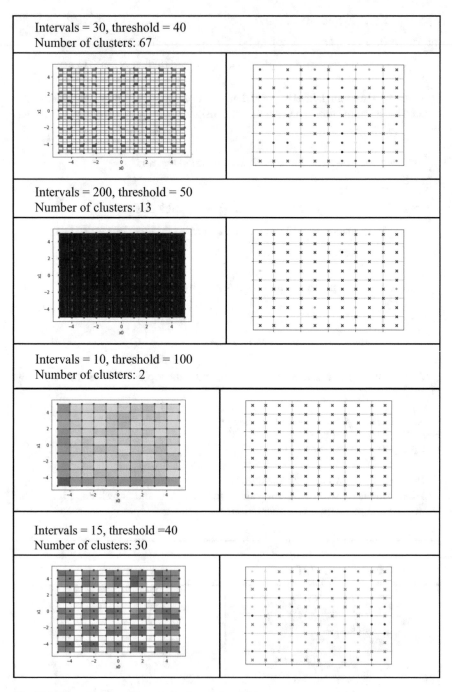

Fig. 16.3 Clustering for various intervals and thresholds

As you observe, clustering largely depends on the values of the two parameters. The source for this project is available in the book's repository.

Summary

In this chapter, you studied two algorithms for clustering large spatial datasets. The STING algorithm can answer the queries without re-scanning the entire dataset and is thus considered query independent. The CLIQUE is a subspace clustering algorithm that uses a bottom-up approach to find clusters. It is considered both density and grid based. It automatically identifies the subspaces of a high-dimensional data space and ideally applies to large-dimensional datasets.

After covering the various clustering algorithms, it is now time to move on to a technology that brought radical changes in machine learning—we call it ANNs (artificial neural networks).

Part III
ANN: Overview

So far, you have learned the classical method (GOFAI) of machine learning. To meet the AI application requirements of today's world, a data scientist must use yet another approach—ANN. The ANN was invented several decades ago, but showed its practical usefulness in recent years. Today's success in ANN technology in solving modern AI problems is mainly because of the availability of high computing and memory resources. In the next two chapters, I will introduce the ANN technology from the standpoint of a data scientist.

Designing an ANN architecture is trivial. You just need to decide upon the number of layers, the number of nodes in each layer, the connection between the layers, and the activation function to fire the neurons. After you design the architecture, decide on the optimization and loss functions for training the network. The number of such functions is small as compared to ones that you learned in GOFAI. Thus, many times working with ANN approaches becomes a lot easier than using the classical approach. You train the model over many epochs until the accuracy score saturates.

Researchers have designed many complicated networks to solve specific problems, such as object detection, creating images, and processing natural language text. I have discussed many such architectures that help in creating AI models on image and text datasets. You will learn CNN, GAN, RNN, LSTM, BERT, and others.

This section also dedicates one full chapter to how a data scientist develops ANN-based applications. I show the application creations on both text and image datasets using many pre-trained models. We call this approach transfer learning. Each project shows how a data scientist would evaluate different approaches on the same dataset to achieve the best-performing machine learning model.

So, let us begin the ANN study.

Chapter 17
Artificial Neural Networks

A Noticeable Evolution in AI

So far in this book, you learned many traditional ML algorithms, with the focus on where to use each one. As you understand, these traditional machine learning algorithms require a deep knowledge of the concepts on which they are based. To derive the optimized results from these algorithms, you are required to fine-tune several parameters. For this, you will need to have a good understanding of the mathematics they are based on. Thus, the whole ML development looks somewhat scary for a beginner.

With the invention of artificial neural networks (ANN), lots of these processes have been largely simplified. Not only that, you will be able to develop AI applications that were not previously feasible using the conventional ML. The dream of self-driving cars would not have been possible without the use of this technology, which has now advanced to what we know as deep learning networks (DNN). We would not have seen personal assistants like Alexa and Siri without DNN. A Google Translate app would have remained very primitive, and Netflix would not have been able to provide us with very appropriate recommendations as to what it does today. The invention of ANN/DNN made all these possible.

AI Evolution

In the early stages of AI, the computer programs provided the reasoning ability, which was more exhaustive and fast as compared to an average human being. As more data became available, it was impossible for human beings to analyze large volumes of data without the aid of computers. Thus, came machine learning. As you have seen, they developed several machine learning algorithms to do predictive analysis using the well-studied statistical techniques. The predictions made by such models were mostly explainable, as we based them on well-defined rules and computations.

P. Sarang, *Thinking Data Science*, The Springer Series in Applied Machine Learning, https://doi.org/10.1007/978-3-031-02363-7_17

Then, comes artificial neural networks. This is where the scientists tried imitating the brain structure in a computer program. They created an artificial neuron in a computer program which has an analogy to a neuron in our brain. They then connected such neurons to form a network which is like a neuron network inside our brain. We call such networks the artificial neural networks (ANNs).

ANNs certainly brought a new revolution in machine learning. The ANN technology solved the problems which were unsolvable using the traditional machine learning algorithms. So, let us try to understand this new technology.

Artificial Neural Networks

Artificial neural networks are a network of artificial neurons assembled in a specific configuration, which can be extremely simple or highly complicated. So, let us first try to understand what I mean by an artificial neuron, and then I will discuss the various configurations in which we arrange them. An artificial neuron is also called a perceptron in technical terms. I will now describe the perceptron.

Perceptron

This is the basic unit of an artificial neural network. It comprises input values, weights, bias, a weighted sum, and the activation function. Schematically, it is shown in Fig. 17.1.

The node takes multiple inputs denoted by $x_1 \ldots x_n$. Each input is multiplied by its respective weight w_i. We sum all such inputs and add a bias to it. We pass the weighted sum through an activation function, the output of which is either 0 or 1. We

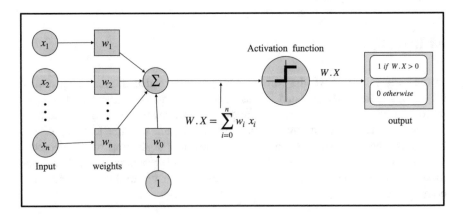

Fig. 17.1 Perceptron schematic

call this phenomenon "neuron is fired or not-fired." The firing or non-firing of a neuron depends on the values of weights w_1 through w_n for the various inputs x_1 through x_n. The assigning appropriate values to these weights for a desired output are called model training.

As a data scientist, you are responsible for the selection of an appropriate activation function and the training, that is, assigning those weights to each neuron in the network. More on this shortly. Let us first look at the ANN architecture.

What Is ANN?

When you arrange the perceptron in a desired configuration, you say that ANN is created. Inspired by biological neural networks, artificial neural network (ANN) is a computing system designed to establish the relationship between a set of data and the expected output. The network architecture is analogous to the neuron network inside our brains. Figure 17.2 shows the schematic of a typical network.

The network has several layers, with each layer containing several nodes. The nodes in the network are interconnected. Note that at each node, we have several connections, and each connection has an associated weight. When we train the network, we find the most appropriate values for all these weights. Looking at the diagram, you easily understand that the total number of weights to be computed is certainly not a small number, and that is why training a neural network is complex and time-consuming.

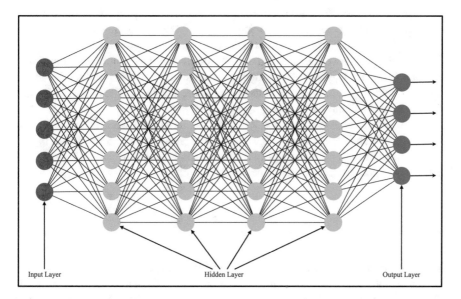

Fig. 17.2 ANN/DNN architecture

Network Training

Just the way we train our brain to do a certain task (we call it learning), we need to train a raw network for performing tasks. This training requires enormous computing resources, depending on the size and complexity of the network. Just the way you keep on teaching the child the same thing again and again until he learns it fully, ANN too learns over several training rounds, called epochs.

The learning improves over many iterations (epochs). You need to show the new and additional information to the network at each iteration. Thus, the amount of data that is required for training a typical network is usually very large—several thousand data points. As a data scientist, this is an important point to note—we can train a classical algorithm with just a few hundred data points, while a neural network typically requires thousands of instances. With the concept of big data, we easily have millions of data points available for ANN training.

Now, let us look at a few basic network architectures.

ANN Architectures

There are primarily three types of architectures:

- Single-layer feedforward network
- Multi-layer feedforward network
- Recurrent network

A single-layer feedforward network, also called single-layer perceptron, contains only the computation neurons of the output layer. As we perform no computation on the input layer, it never gets counted as a layer. This is an acyclic network.

In multi-layer networks, there are one or more hidden layers in between the input and output layers. The input layer supplies input signal to the first hidden layer, the output of which goes as an input to the second one, and so on. The output of the last layer is treated as your final output.

In the case of a recurrent network, it is like a feedforward network, except that it has at least one feedback loop from one of the hidden layers to a previous hidden layer.

With this basic understanding of ANN, let us now look at its advancements and that is DNN and different types of network architectures.

What Is DNN?

When the network architecture becomes really complex with many layers, each having many nodes, we call it a deep neural network (DNN). There is no clear demarcation between ANN and DNN, and many-a-times, people use those terms interchangeably. There is nothing wrong if you too do so.

Our brain can learn multiple things and remember those. However, the DNNs designed so far can do only one kind of task. So, we have different DNN architectures, each specifically designed for a certain task. After a network is trained, it will perform only a certain specific task for which it was trained. For example, a certain network may do language translation, while we train another one for face detection, and so on. Different networks for different tasks.

We have a separate role for designing the network. We call him a machine learning (ML) engineer. The ML engineer, while designing the network, decides upon the number of layers, number of nodes in each layer, the network topology, the activation function, the optimization algorithm, and so on. Though the ML engineer designs different architectures, it is still the job of a data scientist to select the appropriate activation function and the optimization algorithm. I will describe these requirements shortly.

Let us now look at the various configurations largely in practice today.

Network Architectures

Depending on how they connect with each other, you get different network topologies. Each unique architecture has its own name, such as feedforward network, recurrent neural network (RNN), radial basis function neural network, convolutional neural network (CNN), long short-term memory (LSTM) network, and so on. As said earlier, each architecture was designed for a particular purpose and can be trained for solving a specific task.

I will describe these architectures in more detail later in this chapter. As a data scientist, you should understand the purpose for which a particular network was designed and not really worry about its topology. After you understand these various configurations, you will be able to easily create applications based on ANN technology.

The large complex architectures take several hours/weeks of training and also require vast resources. Fortunately, such trained models can be reused by others for further enhancements. We call this transfer learning.

What Are Pre-trained Models?

We now have several ready-to-use pre-trained models. As a data scientist, you can easily extend the functionality of such models, saving you lots of time and resources in designing and training them. You just need to know what models are available. Mainly, the pre-trained models are available that work on image and text data. Applications like object detection and face detection work on image data, while applications such as language translation, news classification, sentiment analysis, and so on work on text data. Toward the end of this chapter, I will describe a few such models from both categories.

With this little background in ANN/DNN, let us understand the key terms used in this space.

Important Terms to Know

I will first describe the various activation functions.

Activation Functions

An activation function in a neural network defines how the weighted sum of the input to a node is transformed into an output. It decides whether a neuron is fired (activated) or not. It brings nonlinearity into the operations done by the various functions in the entire network. You have seen the output of each neuron is given by the equation W.X; without an activation function, it would simply stack the outputs of all layers over one another, resulting in just a linear combination of all. Thus, without an activation function, the entire network is essentially a linear regression model. The activation function helps in keeping the output values restricted to a certain range.

Depending on the type of nonlinearity that you want to bring into the equation, the designers created several kinds of activation functions.

Here are a few widely used activation functions:

- *Sigmoid*—this is traditionally used for binary classification; if $x <= 0.5$ then $y = 0$ or else $y = 1$. This suffers from vanishing gradient problems and is computationally expensive. If the gradients approach close to 0 or 1, the learning rate becomes too slow.
- *Tanh*—this too is used for binary classification; if $x <= 0$, $y = 0$ else $y = 1$. Unlike sigmoid, this is zero-centered, meaning the values range between -1 and $+1$. This too is computationally expensive.
- *ReLU*—Rectified Linear Unit; if $x <= 0$, $y = 0$ else $y = 1$. Thus, it is activated only for positive values. It is computationally less expensive as compared to sigmoid and tanh.
- *Leaky-ReLU*—It solves the dying ReLU problem by adding a small positive slope in the negative plane. This has been widely used in GANs (generative adversarial networks—you will learn more about GANs shortly).
- *Softmax*—This function helps us represent input in terms of a discrete probability distribution. We apply an exponential function to each element of the output layer and normalize their values to ensure that their sum equals 1. We consider the class with the highest confidence score as the final output. We mostly use this as the last layer in multi-class classification problems.

Figure 17.3 presents a table for a few activation functions that gives you a visual presentation, the governing equation, and its derivative.

I have added the source code of the project that was used for generating the above plots in the book's repository for your experimentation.

Name	Plot	Equation	Derivative
Identity		$f(x) = x$	$f'(x) = 1$
Binary Step		$f(x) = \{0\ for\ x < 0;\ 1\ for\ x \geq 0$	$f'(x) = \{0\ for\ x \neq 0$
Logistic		$f(x) = \dfrac{1}{1 + e^{-x}}$	$f'(x) = f(x)(1 - F(x))$
Tanh		$f(x) = tanh(x) = \dfrac{2}{1 + e^{-2x}} - 1$	$f'(x) = 1 - f(x)^2$
ArcTan		$f(x) = tan^{-1}(x)$	$f'(x) = \dfrac{1}{x^2 + 1}$
Rectified Linear Unit (RelU)		$f(x) = \{0\ for\ x < 0;\ x\ for\ x \geq 0$	$f'(x) = \{0\ for\ x < 0;\ 1\ for\ x \geq 0$
Leaky Relu		$f(x) = a = max(0.01x, x)$	$f'(x) = \{0.01\ ;\ if z < 0, 1; otherwise\}$
Parametric Rectified Linear Unit (PReLU)		$f(x) = \{\ \propto x\ for\ x < 0;\ x\ for\ x \geq 0$	$f'(x) = \{\ \propto\ for\ x < 0;\ 1\ for\ x \geq 0$
Exponential Linear Unit		$f(x) = \{\ \propto (e^x - 1)\ for\ x < 0;\ x\ for\ x \geq 0$	$f'(x) = \{\ f(x) + \propto\ for\ x < 0;\ 1\ for\ x \geq 0$
SoftPlus		$f(x) = log_e(1 + e^x)$	$f'(x) = \dfrac{e^x}{1 + e^x}$
Sigmoid		$f(x) = \dfrac{1}{(1 + exp(-x))}$	$f'(x) = f(x) * (1 - f(x))$
Softmax		$f(x) = S = \dfrac{e_i^x}{(\Sigma_{j = \theta}\ e_i^x)}$	$D_j S_i = S_j(\delta_{ij} - S_j)$

Fig. 17.3 Different activation functions

Notice that the table also contains the function's derivative. This is because an activation function must always be differentiable for adjusting the weights during network training. We adjust the weights in a process called back propagation. Let us now try to understand this process.

Back Propagation

When you design a multi-layer network, you assign some initial weights to the various nodes in each layer. Depending on the final output, you adjust the values of these weights and test the output again. You repeat this process multiple times (called epochs) until you are satisfied with your output. Now, how do you know in which direction to adjust the weights? I mean, whether to increase or decrease their values? So, here comes the concept of optimization and loss functions. You use some optimization function that will help in reducing the difference (error) between the current output (predicted) and the desired output (actual). In every iteration, you try to minimize the loss function. You move backward from the output layer to the input layer, making corrections to the weights at each visited layer. We call this back propagation. The optimization function must be differentiable so that you can determine the minima. And that is why I have included the function derivatives in the last column of the activation function table.

The output values can become infinitesimally small or huge because of the optimization operations done at each layer. This can cause the well-known vanishing and exploding gradient problems. So, what are vanishing and exploding gradients?

Vanishing and Exploding Gradients

As you understand now, during training, the gradients are back propagated all the way to the initial layer. At each layer, we do matrix multiplications. For long chains (large number of layers), the gradient will keep shrinking exponentially to very low values and will eventually vanish, in the sense that the network will stop further learning. We call this the vanishing gradient problem. On the other hand, if the gradient moves in the other direction taking large values, it will blow up at some stage to crash the training. We call this the exploding gradient problem.

Next, I will explain the optimization functions.

Optimization Functions

Optimization algorithm is the most important aspect of training a neural network. A loss function, also called an error function, is the objective function for the optimization. The optimization algorithm minimizes the loss function and thus is sometimes referred to as minimization function. Depending on the objective, we use the optimizer for either minimizing or maximizing the objective function. It goes without saying that the loss function needs to be differentiable. Thus, we can say that optimization is a brute-force process for fitting parameters into the loss function to achieve the minimum variance between the predicted and the actual values. In our case of ANN, the weight and the bias become the parameters which need adjustments, and the loss function is some function defined by us with the condition that it should be a continuous, differentiable function.

In neural networks training, we mostly used gradient descent as the optimization strategy. The gradient descent optimizer tries to achieve the optimal values for weights assigned to each node at each layer for which the gradient of the loss function is minimum. Gradient is nothing but the rate of change in the value of some variable. In our case, it is the rate at which the loss function changes with respect to weights. We depict this in Fig. 17.4.

The learning rate defines the incremental step in the above figure. The loss function in the diagram is a continuous convex function. The optimizer goal is to find the global minima or the deepest valley. Note that, in case of non-convex functions, you will observe multiple local minima and maxima.

You must select the appropriate value for the learning rate. If the value is high, you may never reach the local minima as it will keep bouncing back and forth, as seen in Fig. 17.5. If the learning rate is low, you will reach the local minimum after a long time.

Now, let us look at the various optimizers which are used in practice.

Fig. 17.4 Gradient optimization

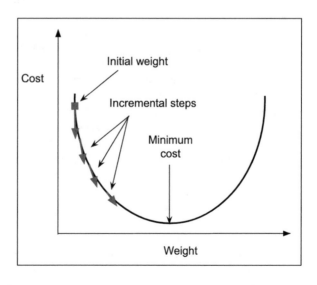

Fig. 17.5 Effect of different
learning rates

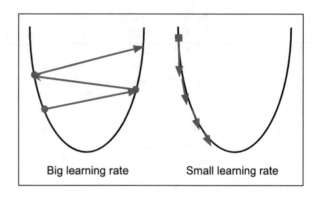

Big learning rate   Small learning rate

Types of Optimizers

I am giving below a list of commonly used optimizers along with their brief
description. As a data scientist, this should suffice in designing your own ANN
architectures. You need not try to understand the mathematics behind these; only the
knowledge of conceptual working would be sufficient.

- Simple gradient descent—Computes the gradient for the entire training dataset.
  Though this is ideally the correct thing to do, the algorithm is computationally
  expensive.
- Stochastic gradient descent (SGD)—Computes the gradient and updates its
  weight for every (x_i, y_i) pair. This is comparatively faster as compared to simple
  gradient descent and is also computationally less expensive. It also carries another
  enormous advantage: that it can update weights with the new data points added on
  the run.
- Mini-batch gradient descent—As the name suggests, the algorithm performs the
  updates on a batch of random points in the dataset. This provides more stable
  updates in the weights compared to stochastic gradient descent. This is also
  computationally less expensive. We send the batch size as a parameter to the
  algorithm. Many modern deep learning libraries use this algorithm.
- Momentum—In case of SGD and mini-batch, the researchers observed fluctuations
  and noticed that it does not follow a steady straight path toward the optimum. To
  remove this noise, the researchers introduced the momentum in the update function.
  The fluctuations were reduced by weight averaging the past parameters.
- Adagrad—In this algorithm, the gradient is adaptive in the sense that for higher
  iterations, the learning rate is reduced.
- RMSProp—This algorithm was developed to overcome the shortcomings of
  Adagrad. It does not decay the learning rate too quickly. RMSProp stands for
  Root Mean Square Propagation and, in most cases, outperforms Adagrad.
- AdaDelta—This stands for Adaptive Delta and removes the shortcomings in
  Adagrad. The delta is the difference between the current weight and newly
  updated weight. The algorithm uses the exponential moving average (EMA) of

square deltas in place of learning rate. Thus, there is no need to explicitly note the learning rate.

- Adam—Adam uses the best of both worlds, momentum and AdaDelta. Its full name is Adaptive Moment Estimation. It uses both the exponential decaying average of past squared gradients as in AdaDelta and also of past gradients as in momentum.

Finally, we come to the loss functions.

Loss Functions

To evaluate the model's training, we use the loss functions. Its value gives us the measure of our training accuracy—how far the predicted value is from the actual. The larger the way, the more training is required; essentially, it means training for more epochs. We use different loss functions for regression and classification.

Regression Loss Functions

I will describe three widely used regression loss functions:

- MAE—mean absolute error; also called L1 loss. The mean absolute error is the average of absolute differences between the predicted and actual values. There is an issue with this loss function—the highly underestimated values cancel equally overestimated values, giving us the wrong idea of the net error.
- MSE—mean squared error, also called L2 loss. Here, we take the average of the squared differences between the predicted and actual values as our error term. Squaring the error helps us resolve the MAE issues. Also, when the input and output values have small scales, the MSE will bring out the error better because of squaring.
- Huber—also called smooth mean absolute error. This is less sensitive to outliers than MSE. It gives out the absolute error. It can amplify the minor errors to quadratic. The hyper-parameter called δ decides this. Huber loss approaches MSE for δ approximating 0 and MAE when δ approximates ∞ (infinity.)

Next, I will describe the loss functions used in classification problems.

Classification Loss Functions

- Binary cross entropy—commonly used for binary classification.
- Hinge—another class entropy function for binary classification. We mainly developed this for SVMs. We use this when the target values are within the dataset.

- Multi-class cross entropy—used for multi-class classifications. Cross entropy computes scores for each class that signify the difference between the predicted and the actual values.
- Sparse multi-class cross entropy—in this case, the output is in the form of one-hot encoded vectors rather than the integer values.

With these preliminary concepts, let us now look at a few commonly used network architectures.

Types of Network Architectures

There are several ANN architectures each developed for a specific purpose. Such architectures typically comprise many nodes and layers. I will now describe the following architectures:

- Convolutional neural networks (CNN)
- Generative adversarial networks (GAN)
- Recurrent neural networks (RNNs)
- Long short-term memory (LSTM)

There are many others, such as restricted Boltzmann machine (RBM), deep belief network (DBN), and self-organizing maps (SOM). The discussions on these would be beyond the scope of this book.

As I said earlier, we developed some of these to work with image data and others on text data. I will discuss the purpose of each as we proceed ahead.

Convolutional Neural Network

Mainly designed for solving computer vision problems, such as image classification and object recognition. The network comprises three types of layers:

- Convolutional
- Pooling
- Fully connected (FC)

The convolutional layer is the first layer, and FC is the last layer. In between, it contains more convolutional and pooling layers. Each convolutional layer identifies some features in the image, progressively increasing in complexity until the entire network finally identifies the intended object. The network heavily uses matrix multiplications while identifying the patterns within an image and thus is computationally expensive.

Convolutional Layer

This is the core building block of CNN. It comprises three components, input data, kernel, and the feature map, as depicted in Fig. 17.6.

We move the kernel or the filter across the fields of an image in order to detect features. This process is called convolution. The kernel is a two-dimensional array of weights and typically has a size of a 3×3 matrix. We compute the dot product between the input pixels and the filter. The filter is then shifted by a stride, repeating the process until they swipe the entire image. We call the output array of all these dot products a feature map, activation map, or a convoluted feature. As the output array does not directly map to input values, we consider this layer as a partially connected layer. You need to decide upon the values of three hyper-parameters before the training begins:

- Number of filters that decide the output depth.
- Stride that decides on the number of pixels that the kernel moves over. Larger the stride, smaller is the output.
- Padding—this decides upon how the output is padded when the filter does not fit the input image.

At the output of a CNN layer, we use ReLU to introduce the nonlinearity. We stack many such convolutional layers in a linear hierarchy. At the output of the last layer, we get the numerical values for the various identified patterns in the image.

Pooling Layer

Pooling layer is used for dimensionality reduction. The pooling can be of two types—max or average. We depict this in Fig. 17.7.

The max pooling takes the maximum value of a pixel covered by the kernel, while the average pooling computes the average of all the pixels covered by the kernel. Though a lot of information is lost because of pooling, it helps in reducing the complexity, improving efficiency, and limiting the risk of over-fitting.

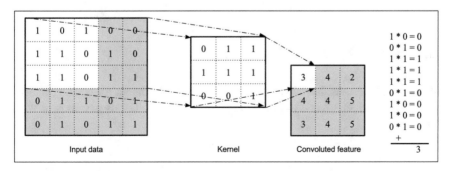

Fig. 17.6 Convolutions

Fig. 17.7 Max and average
pooling

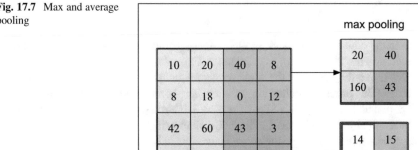

Fully Connected Layer

This layer performs the classification task based on the features extracted from the
entire CNN hierarchy. It uses softmax activation, giving probability values in the
range from 0 to 1, to each class.

CNN Applications

Several CNN variants have been developed, to name a few, AlexNet, VGGNet,
GoogleNet, ResNet, and so on. All of them perform image recognition and computer
vision tasks of deriving meaningful information from images and videos. Based on
this information, you may initiate appropriate actions. Because of this recommendation ability, CNN has been applied in many areas. Social media uses this for
tagging persons in a photograph. Doctors use it to identify malignant tumors. Retail
businesses use it for recommending similar items based on the earlier purchased
(identified) items. Finally, it may also be applied to driverless cars in detecting lanes
and traffic signals.

Generative Adversarial Network

Most likely you must have heard about GAN as an AI model for creating images
and would wonder what it has to do for a data scientist. Let us first look at what
GAN is.

Model Architecture

GAN comprises two neural networks, both CNN based. We call one generator and the other discriminator. An adversarial process trains simultaneously both networks. Generator learns to create images that look real. Discriminator, which is a critic, is trained to tell apart a real image from the fakes. If a generator outperforms the discriminator, we would have the ability to create images that look real.

Let us now look at the two network architectures.

The Generator

The generator architecture is shown in Fig. 17.8.

As you see, it comprises multiple convolutional layers. Starting with a random noise vector of some dimension, we keep upscaling the image through a network of convolution layers until we receive an input of our desired output size (dimension). Each convolution layer, though not shown in the above diagram, is followed by batch normalization and leaky ReLU. Leaky ReLU does not suffer from vanishing gradients and dying ReLU problems.

Now, let us look at discriminator architecture.

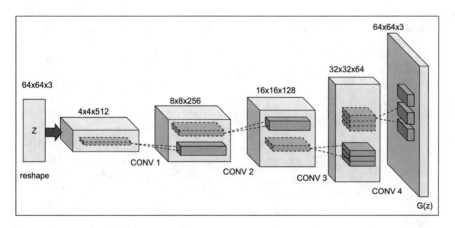

Fig. 17.8 Generator architecture. Image source: Figure 13-2, Apress: Artificial Neural Networks with TensorFlow 2, Author: Poornachandra Sarang

The Discriminator

The discriminator schematics is shown in Fig. 17.9.

Like a generator, the discriminator comprises a series of convolutional layers and downsizes the input image for evaluating it as real or fake.

Let us now look at how the two models work together to achieve our goal of creating images that cannot be shown apart as fake.

How Does GAN Work?

The working schematics of GAN is shown in Fig. 17.10.

GAN training is done in two parts.

In the first part, we keep the generator idle and train only the discriminator. We train the discriminator on actual images over many epochs to make sure that it predicts the real input image (test dataset) as real. After the training, it should be able to tell us if the input image is fake, generated by some other means. How, I will tell you this shortly.

In the second part, we keep the discriminator idle and train only the generator. We start with a random noise vector as an input to the generator and ask it to generate an image of our desired dimension. If the discriminator now tells us that this is a fake image, we retrain the generator for another epoch.

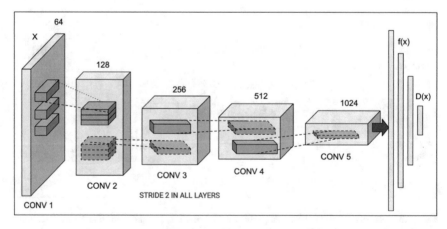

Fig. 17.9 Discriminator architecture. Image source: Figure 13-3, Apress: Artificial Neural Networks with TensorFlow 2, Author: Poornachandra Sarang

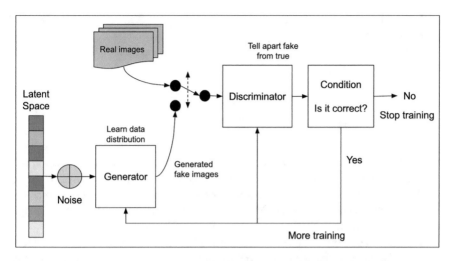

Fig. 17.10 GAN working schematic. Image source: Figure 13-1, Apress: Artificial Neural Networks with TensorFlow 2, Author: Poornachandra Sarang

In fact, we repeat the above steps of training generator and discriminator over many epochs, manually examining the generated image periodically. When the discriminator cannot say that the generated image is fake, we stop the training.

Now, let us look at what is the use of a GAN for a data scientist.

How Data Scientists Use GAN?

The image datasets are always scarce in terms of their data points. You have used MNIST dataset for handwritten digits and also alphabets. Creating many handwritten images is a laborious job. You can use GAN for creating those additional images to increase the dataset size. Note, training a neural network requires a sufficiently large dataset.

Though GAN was designed to create images, we can use it even for creating numerical data points. We had one such requirement in a trivial application. The application was detecting malicious activities (attacks) on a network. Getting the network data for training, testing was an enormous challenge. So, we had to generate our own additional data. The data shape was 60×1, having 60 columns each of numeric type. We use WGAN (Wasserstein GAN) and WGANGP (Wasserstein GAN with Gradient Penalty) to augment our dataset.

These kinds of requirements are rare; however, as a data scientist, you should understand the power and usefulness of GAN.

After studying the architectures designed mainly for image data, let us look at the neural networks designed for text data.

Recurrent Neural Networks (RNN)

Like CNN, RNN was researched decades ago; however, its full potential could not be exploited because of the lack of computational resources in those days. With the availability of resources, they successfully used RNN in a wide variety of domains. We use it in speech-to-text, language translations, time series forecasting, image captioning, question answering, and many other applications.

The major key point to note in RNN is its ability to remember the past. A typical dense neural network has no ability to remember what it learned in the past. In areas like language translations, remembering the previous words is important; otherwise the translation would become a mere word-to-word translation, which will carry a little sense in the translated language. Just for your quick knowledge, Fig. 17.11 shows the RNN architecture.

As seen in the diagram, each node except the first one receives the output of the previous node as an additional input. Thus, the output of each node will depend on its own input, plus the findings of the previous node. Thus, the findings can propagate through many layers until the final output. When you train RNNs for larger depths, you will encounter the vanishing gradient problem. Because of this limitation of RNN, other enhanced architectures soon replaced those.

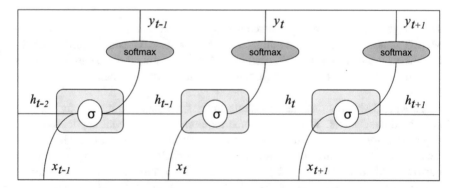

Fig. 17.11 RNN architecture. Image source: Figure 7-2, Apress: Artificial Neural Networks with TensorFlow 2, Author: Poornachandra Sarang

As a data scientist, you need not go deep into RNN working. The model is not in common use these days. However, basic understanding and the concept of remembering the past is important to understand the further architectures.

Long Short-Term Memory (LSTM)

LSTM is a special kind of RNN. It solves the problem of vanishing gradients observed in RNN and can learn long-term dependencies. By its very nature, it has abilities to remember information for longer periods of time. Like RNN, LSTM comprises repeated modules that are much more complicated than you observed in case of RNN. We depict the LSTM module in Fig. 17.12.

Each module comprises four network layers:

- Forget gate
- Input gate
- Update gate
- Output gate

I will now highlight each of these networks in the complete module diagram and briefly explain its purpose.

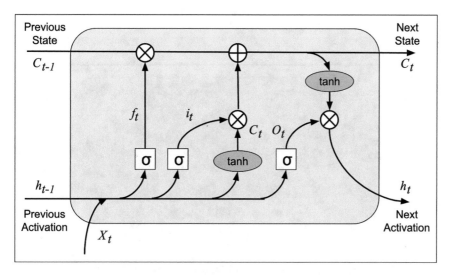

Fig. 17.12 LSTM architecture. Image source: Figure 7-3, Apress: Artificial Neural Networks with TensorFlow 2, Author: Poornachandra Sarang

Forget Gate

The forget gate network in the LSTM module is highlighted in Fig. 17.13.

The forget gate uses sigmoid activation and decides on what information to throw and what to keep. It takes the hidden state of the previous layer as input besides the current input and spills out a binary output. If the output is false, it tells the module to forget the information. Mathematically, it is expressed as:

$$f_t = \sigma\left(W_f.[h_{t-1},x_t] + b_f\right)$$

Input Gate

The input gate is highlighted in Fig. 17.14.

The input gate takes input the current input and the previous hidden state. The sigmoid input layer decides on the values to update, and the tanh activation creates a vector of new candidate values. The two outputs, sigmoid and tanh, are then multiplied. The output contains the important information from the previous module—the time stamp. Mathematically, this is expressed as:

$$i_t = \sigma(W_i.[h_{t-1},x_t] + b_i)$$

$$\widetilde{C}_t = \tan h(W_{c*}[h_{t-1},x_t] + b_c)$$

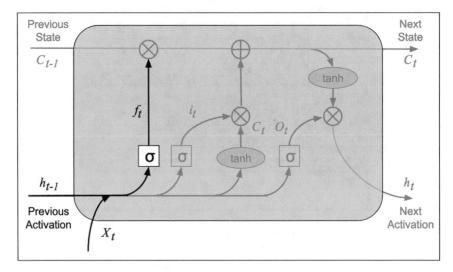

Fig. 17.13 Forget gate. Image source: Figure 7-4, Apress: Artificial Neural Networks with TensorFlow 2, Author: Poornachandra Sarang

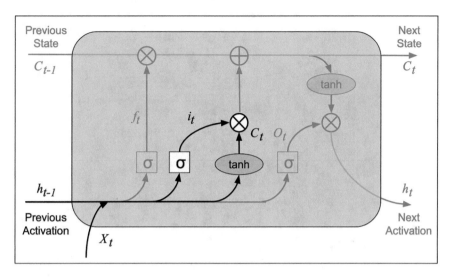

Fig. 17.14 Input gate. Image source: Figure 7-5, Apress: Artificial Neural Networks with TensorFlow 2, Author: Poornachandra Sarang

Update Gate

The update gate is highlighted in the module schematics shown in Fig. 17.15.

Here, the information from the old cell state is updated with the values got from the input gate to create a new cell state. Mathematically, this is expressed as:

$$C_t = f_t * C_t + i_t * \tilde{C}t$$

Output Gate

The output gate is highlighted in Fig. 17.16.

The sigmoid in this network decides upon the parts of the current cell's state that we are going to keep in the output. We multiply together the two outputs to decide the final output state. Mathematically, this is expressed as:

$$o_t = \sigma(W_o[h_{t-1}, x_t] + b_o)$$
$$h_t = o * \tan h(C_t)$$

Thus, the whole LSTM module carries forward the information from the previous state, making the complete chain of modules to remember the information for a long time. The LSTM architecture became quite popular and was successfully applied in many problem domains.

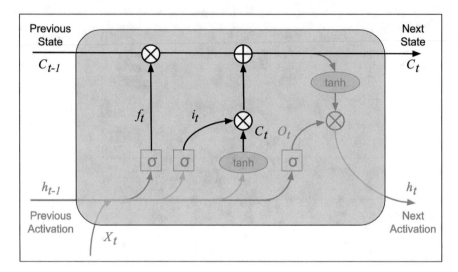

Fig. 17.15 Update gate. Image source: Figure 7-6, Apress: Artificial Neural Networks with TensorFlow 2, Author: Poornachandra Sarang

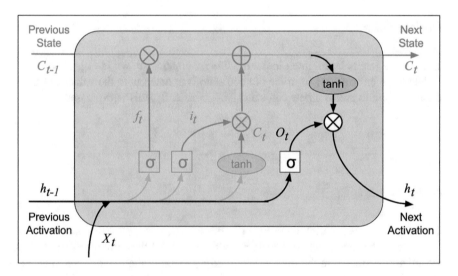

Fig. 17.16 Output gate. Image source: Figure 7-7, Apress: Artificial Neural Networks with TensorFlow 2, Author: Poornachandra Sarang

LSTM Applications

Some of the LSTM applications are:

- Predicting next character/word
- Word/sentence completion

- Learning semantics of a large chunk of text—helps in creating new novels in the style of famous authors like Shakesphere and Agatha Christie
- Neural machine translation

Let us now look at the pre-trained models and the idea of transfer learning.

Transfer Learning

Deep neural networks usually have very complex architectures, require heavy resources for training, take weeks of training times, and consume lots of data in the size of several gigabytes to even terabytes. To meet these needs for developing DNN applications is beyond the scope of an individual data scientist. Only tech-giants could meet such requirements, and yes, they have trained DNNs to solve the problems which a data scientist previously could not solve. Fortunately, these tech giants have made their work available for public use. A data scientist can now extend their models to meet his own requirements. And that is what we call it as transfer learning. There are several models available for your use. They mostly work on text and image data, where the corpus is really huge. I will list out a few such trained models which you can use in your own applications.

Pre-trained Models for Text

I will briefly introduce you to the following five pre-trained models.

- Word2Vec
- Glove
- Transformers
- BERT
- GPT

Each one was designed for a specific purpose and finds its use in several ML applications that use text datasets.

Word2Vec

Consider developing a ML model to predict the next word or finding a missing word in a sentence. Developing such models requires a good vocabulary, which helps in capturing context of a word in a document, discovering semantic/syntactic similar-ities between words, relation with other words in the document, and so on. As you can imagine, a vocabulary can be extremely large even for a small- to medium-size document. Representing the entire vocabulary in a machine-readable format, i.e., binary format, would require prodigious memory. The embeddings provide a compact representation of the vocabulary. The Word2Vec is a method to construct such an embedding. The embedding can be constructed using two methods—Skip Gram and Common Bag Of Words (CBOW). Many large-sized pre-trained Word2Vec models are available for developing various kinds of NLP applications. You will even find Sentence2Vec models that compute sentence similarities.

Glove

Glove, an open-source project launched in 2014 at Stanford, is an unsupervised machine learning algorithm that creates vector representations for words. The algorithm discovers the semantic similarities between the words by giving ratios of word-to-word co-occurrence probabilities. For example, the probability of the word "cream" occurring next to "ice" would be very high. The algorithm was applied on a huge text corpus such as Wikipedia, Common Crawl, and Twitter, and the embeddings for dimensions of 50, 100, 200, and 300 were developed. You can use these embeddings into your NLP projects to build your own dictionaries.

Transformer

This is the latest innovation in natural language modeling. This has virtually replaced LSTMs—the model used popularly for developing NLP applications. It is based on the famous paper "Attention Is All You Need" and has drastically improved the performance of neural machine translation applications.

BERT

BERT stands for Bidirectional Encoder Representations from Transformers. Jacob Delvin et al. introduced this model in 2018 in their paper titled "BERT: Pretraining of Deep Bidirectional Transformers for Language Understanding." You can easily use a pre-trained BERT model in your NLP applications—just adding one layer is all it takes to do. You can easily create highly accurate applications of language translation, question-answering, and so on.

GPT

This is an autoregressive language model that produces human-like text. Developed at OpenAI, it has given an outstanding performance in text generation. It is so powerful in text generation that one can easily create fake news that can pass as real ones. Because of this and other malicious applications that could possibly be created using GPT, OpenAI has released only a small model for experimentation.

Unlike BERT, GPT is unidirectional and was trained on 175 billion parameters. The current version as of this writing is GPT-3, and new versions are periodically introduced.

With this brief on the different pre-trained models for text data, in the next chapter, I will show you how to use those in your NLP applications.

I will now similarly give you a brief on some of the pre-trained models developed for image-based applications.

Pre-trained Models for Image Data

The pioneer in this field is probably AlexNet, a CNN architecture that achieved a top 5 position in the ImageNet Large Scale Visual Recognition Challenge in September 2012. The ImageNet, the first of its kind in terms of scale, comprises over 14 million images labeled into 27 sub-trees and 21,841 subcategories. They developed many machine learning models to classify this dataset. Two years later, Google DeepMind and the University of Oxford launched VGG16 and VGG19. Following this, we have a family of architectures in this category including ResNet, Inception, Xception, MobileNet, DenseNet, EfficientNet, and hybrids. I will briefly introduce you to a few models in this category.

Residual network (ResNet) introduced in 2015 by Shaoqing et al. is one of the popular and successful deep learning models. They built it on constructs taken from pyramidal cells in the cerebral cortex. It comprises 50, 101, or 151 layers. They use skip connections to jump over some layers. It provides a high accuracy when compared to other networks in its category. It takes a long time to train and infer.

VGG 16 is a CNN model from researchers at the University of Oxford that achieved 92.7% top 5 test accuracy in ImageNet. They trained it for several weeks using the NVIDIA Titan Black GPUs.

Inception V3 attained greater than 78.1% accuracy on ImageNet. The Xception slightly outperformed Inception V3 on ImageNet, but showed a significant improvement over Inception V3 on larger datasets comprising 350 million images and 17,000 classes. DenseNet is yet another CNN architecture that reached state-of-the-art results on ImageNet. The EfficientNet improves accuracy and efficiency by using AutoML and employing model scaling.

MobileNet was designed with mobile devices in mind for image classification, object detection, and more.

YOLO (You Only Look Once) is a state-of-the art, real-time object detection system that draws bounding boxes around the detected objects and predicts its class with the corresponding probabilities. Actually, it is an algorithm that uses ANN to perform real-time object detection. It is quite fast and accurate. We use it in a wide variety of applications like detecting traffic signals, parking meters, and so on. If you come across a requirement for analyzing the traffic, say how many heavy vehicles are moving toward east at a particular junction, think of using YOLO.

Finally, to say, as a data scientist, you have several options of pre-trained models for developing your image-based ML models.

Now, let us look at some advantages and disadvantages in using the ANN technology over the classical ML—GOFAI.

Advantages/Disadvantages

Here are some of the advantages of using ANN technology:

- The artificial neural networks can easily learn even nonlinear and complex relationships between the input and output.
- It can handle huge datasets, which is a severe limitation of GOFAI.
- We can distribute the training across parallel processors, which reduces the training times.
- Once trained, it also has an ability to generalize, i.e., it can infer new unseen relationships.
- One major advantage is that the features engineering is built in. During training, the features are automatically extracted.
      Here are a few disadvantages:
- Computationally more expensive than GOFAI.
- Requires much larger datasets as compared to those required in case of classical ML.
- The network design is critical and may take a long time to achieve the desired results. (These days you have tools similar to AutoML that aid in designing efficient networks.)
- Explainability—a neural network functions like a black box. You do not know why it came up with a certain output.

Summary

Introduction of ANN technology has definitely brought a radical change in machine learning. Though the technology is many decades old, its success came only with the availability of computing resources that we have today. While developing the machine learning models, the data scientists now have a choice between the two technologies—classical statistical-based versus artificial neural networks-based. The ANN/DNN technology requires heavy resources, huge datasets, and several hours/days/weeks of training times, while the traditional classical ML technology works superbly for small datasets. The ANN technology fails totally if the datasets are too small. So, this can become a deciding factor while selecting the technology for model development.

In this chapter, you learned this new ANN technology. You learned many basic terms used in this technology, such as activation functions, optimization, and loss functions. As a data scientist, the selection of these functions plays a vital role in the success of your project. You learned many popular network architectures and their purpose. You also learned about pre-trained networks and transfer learning for model development.

In the next chapter, I will describe two applications—one for image data and the other for text data. This will give you some overview of what kind of preprocessing is required for such data types. I will also use the pre-trained models in these applications to illustrate their use.

Chapter 18
ANN-Based Applications

Text and Image Dataset Processing for ANN Applications

Having studied the ANN and the different network configurations of ANN, this chapter now focuses on developing projects based on ANN technology. The purpose is to show you a data scientist's approach to model building using ANN technology.

I will present projects based on both text and image data, so that you will understand how to process such data before feeding it to a neural network. I will show you how to develop your ANN architecture as well as how to use a pre-trained model.

First, I will discuss a project based on text data.

Developing NLP Applications

In this project, you will classify the news headlines under several pre-defined topics. You will learn three approaches—transformers, bag-of-words, and pre-trained word embeddings. In the first part of the project, you will first do some exploratory data analysis and cleansing, followed by a wide variety of methods for data preparation, as required by these NLP models. We will use the dataset from UCI repository comprising 93,000+ news items classified into four different categories—Economy, Obama, Microsoft, and Palestine. In the second part, I will show you the development and applications of the various NLP models.

Dataset

After loading the dataset into Pandas dataframe, I printed its info which can be seen in Fig. 18.1.

Out of these, we will use only the Headline and Topic columns. Figure 18.2 shows the data distribution.

© The Author(s), under exclusive license to Springer Nature Switzerland AG 2023 289
P. Sarang, *Thinking Data Science*, The Springer Series in Applied Machine Learning,
https://doi.org/10.1007/978-3-031-02363-7_18

```
RangeIndex: 93239 entries, 0 to 93238
Data columns (total 11 columns):
 #    Column              Non-Null Count    Dtype
---   ------              --------------    -----
 0    IDLink              93239 non-null    float64
 1    Title               93239 non-null    object
 2    Headline            93224 non-null    object
 3    Source              92960 non-null    object
 4    Topic               93239 non-null    object
 5    PublishDate         93239 non-null    object
 6    SentimentTitle      93239 non-null    float64
 7    SentimentHeadline   93239 non-null    float64
 8    Facebook            93239 non-null    int64
 9    GooglePlus          93239 non-null    int64
 10   LinkedIn            93239 non-null    int64
```

Fig. 18.1 Dataset information

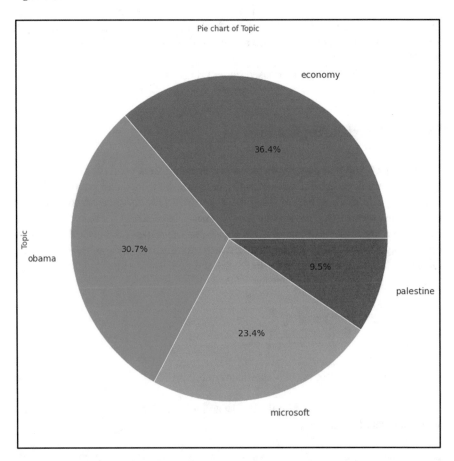

Fig. 18.2 Data distribution for four classes

This shows the dataset is well-balanced. I also checked for the null values in the dataset.

Text Preprocessing

The headline text may contain some URLs. To remove the URLs, I define the following function and applied it to dataset as follows:

```
def remove_URL(text):
    url = re.compile(r'https?://\S+|www\.\S+')
    return url.sub(r'',text)


df['Headline']=df['Headline'].apply(
                        lambda x : remove_URL(x))
```

The HTML tags in the text were removed using the following function:

```
def remove_html(text):
    html=re.compile(r'<.*?>')
    return html.sub(r'',text)


df['Headline']=df['Headline'].apply(
                        lambda x : remove_html(x))
```

The emojis were removed using the following code:

```
def remove_emoji(text):
    emoji_pattern = re.compile("["
            u"\U0001F600-\U0001F64F"
                # emoticons
            u"\U0001F300-\U0001F5FF"
                # symbols & pictographs
            u"\U0001F680-\U0001F6FF"
                # transport & map symbols
            u"\U0001F1E0-\U0001F1FF"
                # flags (iOS)
            u"\U00002702-\U000027B0"
            u"\U000024C2-\U0001F251"
            "]+", flags=re.UNICODE)
    return emoji_pattern.sub(r'', text)

df['Headline']=df['Headline'].apply(
                    lambda x: remove_emoji(x))
```

The punctuations were removed with the following code:

```
def remove_punct(text):
    table=str.maketrans('','',string.punctuation)
    return text.translate(table)

df['Headline']=df['Headline'].apply(
                    lambda x : remove_punct(x))
```

The numbers were removed using the following statement:

```
df['Headline']=df['Headline'].str.replace('\d+', '')
```

The multiple spaces are removed with this code:

```
df['Headline']=df['Headline'].str.replace('     ', ' ')
df['Headline']=df['Headline'].str.replace('       ', ' ')
df['Headline']=df['Headline'].str.replace(
                              '\xa0 \xa0 \xa0', ' ')
df['Headline']=df['Headline'].str.replace('  ', ' ')
df['Headline']=df['Headline'].str.replace('—', ' ')
df['Headline']=df['Headline'].str.replace('-', ' ')
```

We did this entire above text preprocessing to reduce the vocabulary size. For large corpus, the vocabulary is usually very large and for NLP applications; it becomes absolutely necessary to create trainable ML models.

Next, I will show you how to apply the pre-trained BERT model on this cleansed data for news classification.

Using BERT

The Huggingface library provides the ready-to-use implementation of BERT. You install the library using the pip command:

```
!pip install transformers
```

Then, you import the BERT model into your project.

```
from transformers import TFBertModel, BertConfig,
                          BertTokenizerFast
```

Creating Training/Testing Datasets

Our target column is categorical. We convert this to numeric using the following code:

```
data = df[['Headline', 'Topic']]
data['Topic_label'] = pd.Categorical(data['Topic'])
data['Topic'] = data['Topic_label'].cat.codes
```

We split the entire dataset into training/testing in the ratio 90:10.

```
data_train, data_test = train_test_split(
                            data, test_size = 0.1)
```

Setting Up BERT

We load the BERT model called "bert-base-uncased" and set up its configuration using the following code:

```
# Name of the BERT model to use
model_name = 'bert-base-uncased'

# Max length of tokens
max_length = 45

# Load transformers config and set output_hidden_states
to False
config = BertConfig.from_pretrained(model_name)
config.output_hidden_states = False
```

We load the tokenizer:

```
tokenizer = BertTokenizerFast.from_pretrained(
                pretrained_model_name_or_path =
                model_name, config = config)
```

Finally, we load the model:

```
transformer_model = TFBertModel.from_pretrained(
                    model_name, config = config)
```

Model Building

We now design our ANN and incorporate the above-loaded pre-trained BERT model in it. First, you need to set up the input layer to BERT with the following code:

```
bert = transformer_model.layers[0]

# Build your model input
input_ids = Input(shape=(max_length,),
                    name='input_ids', dtype='int32')
inputs = {'input_ids': input_ids}
bert_model = bert(inputs)[1]
```

We then add a dropout layer to the network to reduce over-fitting.

```
dropout = Dropout(config.hidden_dropout_prob,
                    name='pooled_output')
pooled_output = dropout(bert_model, training=False)
```

We add a dense layer with the number of layers equal to the number of classes in our label. This is the output layer.

```
topics = Dense(units=len(
            data_train.Topic_label.value_counts()),
            kernel_initializer=TruncatedNormal(
            stddev=config.initializer_range),
            name='topic')(pooled_output)
outputs = {'topic': topics}
```

Finally, we assemble our model using the above layers:

```
model = Model(inputs=inputs, outputs=outputs,
                name='BERT_MultiLabel_MultiClass')
```

At this point, our model definition is complete. The model summary can be seen in Fig. 18.3.

We will now train the model.

Model Training

For training the model we need to set up the optimizer.

```
Model: "BERT_MultiLabel_MultiClass"
_____
Layer (type)                Output Shape            Param #
====================================================================
input_ids (InputLayer)      [(None, 45)]            0

bert (TFBertMainLayer)      TFBaseModelOutputWithPoo 109482240
                            lingAndCrossAttentions(l
                            ast_hidden_state=(None,
                            45, 768),
                             pooler_output=(None, 76
                            8),
                             past_key_values=None, h
                            idden_states=None, atten
                            tions=None, cross_attent
                            ions=None)

pooled_output (Dropout)     (None, 768)             0

topic (Dense)               (None, 4)               3076

====================================================================
Total params: 109,485,316
Trainable params: 109,485,316
Non-trainable params: 0
_____
```

Fig. 18.3 Model summary

```
optimizer = Adam(learning_rate=5e-05,epsilon=1e-08,
                    decay=0.01,clipnorm=1.0)
```

I used Adam optimizer with a few parameter settings. We set the loss function to *CategoricalCrossentropy*:

```
loss = {'topic': CategoricalCrossentropy(
                            from_logits = True)}
```

We compile the model using these optimizer and loss values.

```
model.compile(optimizer = optimizer, loss = loss,
                        metrics = ['accuracy'])
```

We ready the output for the model:

```
y_topic = to_categorical(data_train['Topic'])
```

For feeding data into the model, we must first tokenize it. We set up the tokenizer as follows:

```
x = tokenizer(
    text=data_train['Headline'].to_list(),
    add_special_tokens=True,
    max_length=max_length,
    truncation=True,
    padding=True,
    return_tensors='tf',
    return_token_type_ids = False,
    return_attention_mask = True,
    verbose = True)
```

Finally, we start the training by calling the *fit* method:

```
history = model.fit(
    x={'input_ids': x['input_ids']},
    y={'topic': y_topic},
    validation_split=0.1,
    batch_size=64,
    epochs=2,
    verbose=1)
```

Note that I have set up the epochs to just a small value. The training may take a noticeable time to complete.

Figure 18.4 shows the accuracy and the loss metrics.

As you can see, the model has reached an accuracy of more than 98% in just two epochs. This is a big advantage in using the pre-trained BERT model.

Model Evaluation

We now evaluate the model's performance on our test dataset. For this, we need to tokenize the test dataset as we did for the training one.

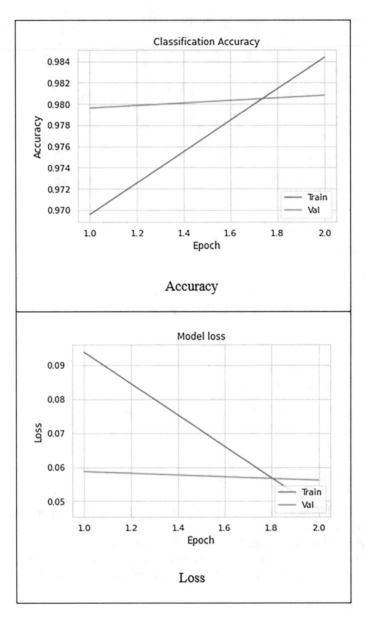

Fig. 18.4 Accuracy/loss metrics

```
test_x = tokenizer(
    text=data_test['Headline'].to_list(),
    add_special_tokens=True,
    max_length=max_length,
    truncation=True,
    padding=True,
    return_tensors='tf',
    return_token_type_ids = False,
    return_attention_mask = False,
    verbose = True)
```

We evaluate the performance by calling the model's *evaluate* method.

```
model_eval = model.evaluate(
    x={'input_ids': test_x['input_ids']},
    y={'product': test_y_topic}
)
```

I got an accuracy of 97.88% in my test run.
You do the inference on the test dataset by calling the model's *predict* method.

```
label_predicted = model.predict(
    x={'input_ids': test_x['input_ids']},
)
```

The output is a matrix of predictions where the highest value in each row will specify the predicted class.

Figure 18.5 shows the classification report generated in my test run.

	precision	recall	f1-score	support
0	0.98	0.98	0.98	3369
1	0.99	0.99	0.99	2167
2	0.98	0.98	0.98	2923
3	0.96	0.98	0.97	864
accuracy			0.98	9323
macro avg	0.98	0.98	0.98	9323
weighted avg	0.98	0.98	0.98	9323

Fig. 18.5 Classification report

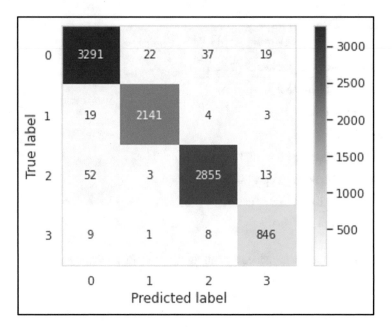

Fig. 18.6 Confusion matrix

Figure 18.6 shows the confusion matrix.

From the above discussions, you can see how easy it is to use a pre-trained model for your NLP applications. That, however, does not make the old technologies absolutely obsolete. For trivial tasks, data scientists still use these old technologies to avoid the overheads like resources and training times in using the new technology.

Using Embeddings

In this part of the project, we will use bag-of-words and the pre-trained word embeddings. We need to remove stop words from our text corpus. For this, I will download the stop words for English language from the *nltk* toolkit.

```
import nltk
nltk.download("stopwords")
from nltk.corpus import stopwords
stop_words = set(stopwords.words("english"))
```

We will first do some analysis on our dataset to understand the frequency of words under each topic. I created a bar plot that displays the ten most common words in the Economy topic using the following code:

```
word_count = Counter(" ".join(
                df[df['Topic']=='economy']
                ['Headline']).split()).most_common(100)
x=[]
y=[]
for word,count in word_count:
    if (word.casefold() not in stop_words) :
        x.append(word)
        y.append(count)


sns.barplot(x=y[:10],y=x[:10])
plt.title('10 most common words in Economy Topic')
```

Such bar plots for all four topics are shown in Fig. 18.7.

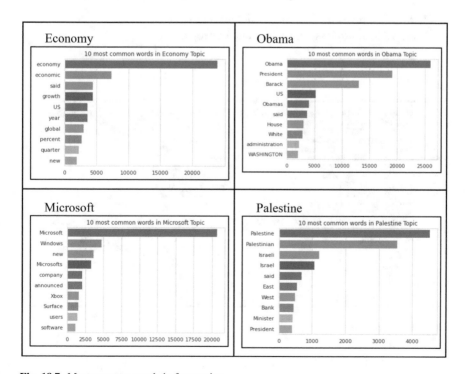

Fig. 18.7 Most common words in four topics

This visualization gives you the idea of the most important words to look out for in a given topic. We will also do an n-gram analysis to understand the adjoining word sequences.

N-gram Analysis

We write a function to detect n-grams:

```
def generate_ngrams(text, n_gram):
    token = [token for token in text.lower().split(' ')
                if token != '' if token not in stop_words]
    ngrams = zip(*[token[i:] for i in range(n_gram)])
    return [' '.join(ngram) for ngram in ngrams]
```

For generating bi-grams under the topic Economy, we define a trivial *while* loop:

```
for instance in df[df['Topic']=='economy']['Headline']:
    for word in generate_ngrams(instance, n_gram=2):
        economy_bigrams[word] += 1
```

Figure 18.8 shows the bi-grams under all four topics.
Figure 18.9 shows tri-grams under the four topics.

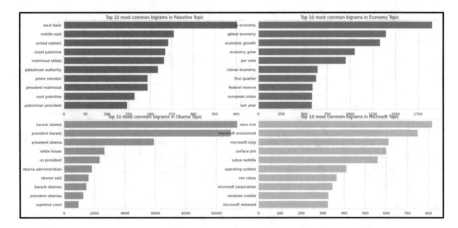

Fig. 18.8 Bi-grams in four topics

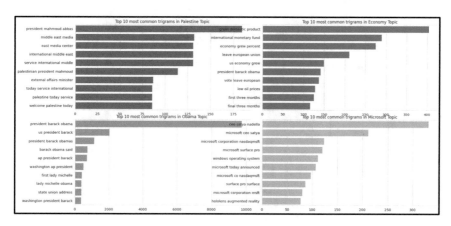

Fig. 18.9 Tri-grams in four topics

This kind of exploratory analysis is required while using your own network models. Note that in BERT and other pre-trained models, these things are taken care of internally.

Next, we tokenize the text corpus.

Tokenizing

We use the tokenizer provided in the TensorFlow library that now fully integrates Keras.

```
import tensorflow as tf
from tensorflow.keras.preprocessing.text
      import Tokenizer
from tensorflow.keras.preprocessing.sequence
      import pad_sequences
```

Remove Stop Words

```
df['Headline_without_stopwords'] =
          df['Headline'].ap-ply(lambda x: ' '.join(
          [word for word in x.split() if word not in
          (stop_words)]))
```

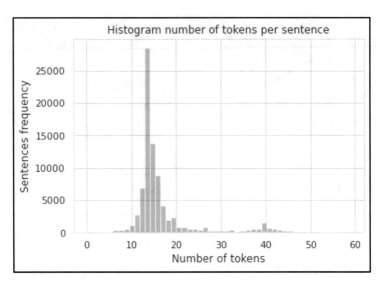

Fig. 18.10 Frequency of tokens

We create the training and testing datasets as earlier and then apply the tokenizer:

```
tokenizer = Tokenizer(num_words=vocab_size, oov_token=oov_tok)
tokenizer.fit_on_texts(train_sentences)
```

Next, we have to get the length of each sequence to determine the optimal size for the sequence length. Figure 18.10 shows the distribution for various lengths.

From the above plot, we can conclude our sequence length to be 45. We now need to pad all our training sequences to this length.

```
max_length = 45
trunc_type = 'post'
padding_type = 'post'


train_padded = pad_sequences(train_sequences,maxlen=max_length,
                  padding=padding_type,truncating=trunc_type)
```

Likewise, you will also pad the validation tokens. Finally, you will apply the tokenizer and create the training and validation sets for inputting to our models.

Note that we had to do a lot of data preparation, which was automatically taken care of while using BERT.

We are now ready to create our models.

Model Building

We will create four different models. In the first model, we will use our own previously created embeddings—bag-of-words. In the rest three, we will use *Glove* embeddings of 100 dimensions. We will simply try out the different network configurations to study how they affect the model's performance.

Using Own Embeddings: Model 0

We define our network model:

```
embedding_dim = 32
model = tf.keras.Sequential([
            tf.keras.layers.Embedding(vocab_size,
                embedding_dim, input_length=max_length),
            tf.keras.layers.GlobalAveragePooling1D(),
            tf.keras.layers.Dense(24, activation='relu'),
            tf.keras.layers.Dense(5, activation='softmax')
])
```

Note, we have set the embedding dimensions to 32 for our vocabulary. Using Adam optimizer and *sparse_categorial_crossentropy* as our loss function, we compile the model.

```
opt=Adam(learning_rate=5e-3)
model.compile(loss='sparse_categorical_crossentropy',
                optimizer=opt, metrics=['accuracy'])
```

Figure 18.11 shows the model summary.

The first layer in the network is our embedding layer created using the embedding class of Keras. The second layer is the global average pooling, followed by a 24-dimension dense layer. The last layer is a dense *softmax* activation layer for outputting five classes.

You train the model using the following code:

```
Model: "sequential"
_____
 Layer (type)                Output Shape              Param #
=================================================================
 embedding (Embedding)       (None, 45, 32)            960000

 global_average_pooling1d (G  (None, 32)               0
 lobalAveragePooling1D)

 dense (Dense)               (None, 24)                792

 dense_1 (Dense)             (None, 5)                 125

=================================================================
Total params: 960,917
Trainable params: 960,917
Non-trainable params: 0
_____
```

Fig. 18.11 Model summary

```
num_epochs = 10
history = model.fit(train_padded,training_label_seq,
                    epochs=num_epochs,
                    validation_data=(
                    validation_padded,validation_label_seq),
                    verbose=1)
```

As you see in the verbose output, the training time for each epoch is just a few seconds. Compare this with our previous BERT model, which took almost an hour for just two epochs. This model also shows the validation accuracy of 96% in just ten epochs. Figure 18.12 shows the plot of training/validation accuracy and the loss against the number of epochs.

Notice that the acceptable accuracy is reached in about 5–6 epochs, resulting in considerable saving in training time and resources.

As earlier, you can do the predictions and print the classification report and the confusion matrix. Figure 18.13 shows the classification report:

We will now try the Glove embeddings. For this, first we need to construct the weight matrix for the embedding layer.

Embedding Weight Matrix

We will first tokenize the full corpus. In the last model, I had used only 30,000 sentences. The full corpus contains 57,515 sentences. The tokenization code is similar to the earlier one, except for a different set of variables, so I am not including it here. We now download the weights of Glove-100 dim directly from Stanford site:

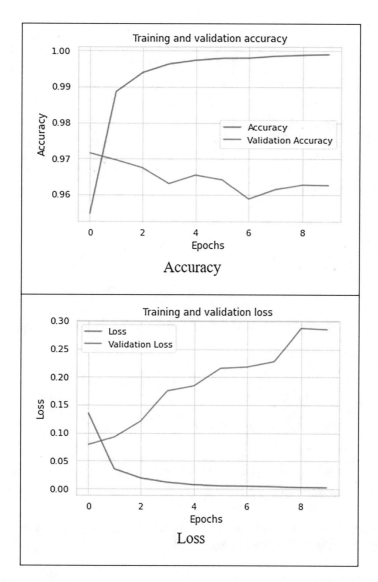

Fig. 18.12 Accuracy/loss metrics

```
!wget "https://nlp.stanford.edu/data/glove.6B.zip"
      -O "glove.6b.zip"
```

When you unzip the downloaded file, you will find the weights for embeddings for four different dimensions—50, 100, 200, and 300. The zip contains the following four txt files:

	precision	recall	f1-score	support
1	0.96	0.95	0.96	3395
2	0.97	0.94	0.96	2989
3	0.97	0.99	0.98	2250
4	0.85	0.99	0.92	689
accuracy			0.96	9323
macro avg	0.94	0.97	0.95	9323
weighted avg	0.96	0.96	0.96	9323

Fig. 18.13 Classification report

```
Archive: glove.6B.zip
 inflating: glove.6B.50d.txt
 inflating: glove.6B.100d.txt
 inflating: glove.6B.200d.txt
 inflating: glove.6B.300d.txt
```

You load the coefficients from the 100-dim file using the following code:

```
embeddings_index = {};
with open('/content/glove.6B.100d.txt') as f:
    for line in f:
        values = line.split();
        word = values[0];
        coefs = np.asarray(values[1:], dtype='float32');
        embeddings_index[word] = coefs;
```

Finally, create the weight matrix for the first embedding layer of our matrix having dimensions of our vocabulary × 100.

```
embeddings_matrix = np.zeros(
                    (vocab_size_glove+1,100));
for word, i in word_index.items():
    embedding_vector = embeddings_index.get(word);
    if embedding_vector is not None:
        embeddings_matrix[i] = embedding_vector;
```

We will now build three different architectures containing the Glove embedding layer.

Glove: Model 1

We define the model:

```
embedding_dim_glove=100
model_glove = tf.keras.Sequential([
    tf.keras.layers.Embedding(vocab_size_glove+1,
            embedding_dim_glove, input_length=max_length,
            weights=[embeddings_matrix], trainable=False),
    tf.keras.layers.GlobalAveragePooling1D(),
    tf.keras.layers.Dense(24,activation='relu'),
    tf.keras.layers.Dense(5,activation='softmax')
])
```

Note that the embedding dimensions are 100. The three remaining layers are the same as the earlier architecture of Model 0. We compile the model as before:

```
opt=Adam(learning_rate=5e-3)
model_glove.compile(loss='sparse_categorical_crossentropy',
                    optimizer=opt,metrics=['accuracy'])
```

Figure 18.14 shows the model summary.

```
Model: "sequential_1"

 Layer (type)                    Output Shape            Param #
=================================================================
 embedding_1 (Embedding)         (None, 45, 100)         5751500

 global_average_pooling1d_1      (None, 100)             0
 (GlobalAveragePooling1D)

 dense_2 (Dense)                 (None, 24)              2424

 dense_3 (Dense)                 (None, 5)               125

=================================================================
Total params: 5,754,049
Trainable params: 2,549
Non-trainable params: 5,751,500
```

Fig. 18.14 Model summary

Note that the model contains more than 5.7+ million parameters out of which only 2549 require training. Compare this with earlier Model 0 which has slightly less than 1 million parameters to be trained.

You now train the model over ten epochs:

```
num_epochs = 10
history_glove = model_glove.fit(
                        train_glove_padded,
                        training_label_seq,
                        epochs=num_epochs,
                        validation_data=(
                        val_glove_padded,
                        validation_label_seq),
                        verbose=1)
```

Each epoch took about 10 s in my run—a noticeable reduction in training time due to the use of a pre-trained model. Unfortunately, the accuracy on validation set was just about 77%, certainly very low as compared to Model 0 and not acceptable to us. So, we will now try changing the network configuration.

Glove: Model 2

This time we will use a SpatialDropout1D layer which is an equivalent of dropout in dense layers applied to 1-D sequences, followed by an LSTM of 45 units—the same as the sentence length. The final layer will be a dense softmax layer as before for classification into five classes. Figure 18.15 shows network summary:

I trained this model for 5 s; each epoch took just about 60 ms for training and achieved a validation accuracy of about 93% toward the end. We will now try one more architecture.

Glove: Model 3

As we observed from Model 2, the LSTM has given us better accuracies; we will go back to our architecture of Model 1. This time we will simply change the global average pooling layer with LSTM. Figure 18.16 shows the new architecture.

This model achieved an accuracy of about 94% in just five epochs. Figure 18.17 shows the classification report for this final model.

Figure 18.18 shows the corresponding confusion matrix.

I will now summarize our entire experimentation.

```
Model: "sequential_2"
_____
Layer (type)                 Output Shape              Param #
=================================================================
embedding_2 (Embedding)      (None, 45, 100)           5751500

spatial_dropout1d (SpatialD  (None, 45, 100)           0
ropout1D)

lstm (LSTM)                  (None, 45)                26280

dense_4 (Dense)              (None, 5)                 230

=================================================================
Total params: 5,778,010
Trainable params: 26,510
Non-trainable params: 5,751,500
_____
```

Fig. 18.15 Network summary

```
Model: "sequential_8"
_____
Layer (type)                 Output Shape              Param #
=================================================================
embedding_8 (Embedding)      (None, 45, 100)           5751500

lstm_5 (LSTM)                (None, 45)                26280

dense_11 (Dense)             (None, 24)                1104

dense_12 (Dense)             (None, 5)                 125

=================================================================
Total params: 5,779,009
Trainable params: 27,509
Non-trainable params: 5,751,500
_____
```

Fig. 18.16 The revised architecture

	precision	recall	f1-score	support
1	0.94	0.93	0.94	3406
2	0.96	0.91	0.93	3054
3	0.95	1.00	0.97	2174
4	0.82	0.95	0.88	689
accuracy			0.94	9323
macro avg	0.92	0.95	0.93	9323
weighted avg	0.94	0.94	0.94	9323

Fig. 18.17 Classification report for the final model

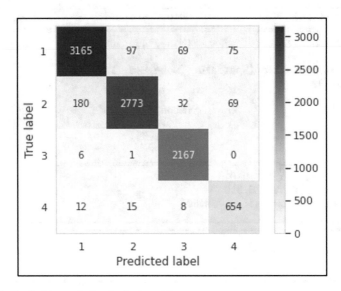

Fig. 18.18   Confusion matrix for the final model

Final Thoughts

From the above observations on the four different models, we conclude that our own embeddings did slightly outperform Glove. We can improve the latter by using higher dimension embeddings and trying out different network configurations. Second, our own embeddings are based on the same dataset, which was once again used for testing. So the vocabulary, while inferring, was known, resulting in better performance. However, you need to take into consideration the training times, resources, and the efforts required in doing EDA while building your own embeddings. Finally, compare all these four models based on embeddings with BERT. The advantages and performance offered by the use of pre-trained models certainly outweigh the others.

The source for this project is available in the book's repository.

Having seen how to develop NLP applications based on text data, I will now discuss the next project that uses image data.

Developing Image-Based Applications

The text data is abundantly available on the web. You have tweets, news feeds, and so on for even real-time data feeds. However, getting a good image dataset for model training is usually a challenge. Facebook and others have released some good datasets for model training. For this project, I am going to use the dataset provided in the TensorFlow library. It is called CIFAR-10, a collection of 60,000 color images. It is a labeled dataset comprising 10 classes, with 6000 images per class. It is split nicely into 50,000 training images and 10,000 testing images. Each image is of dimensions 32×32 pixels. The dataset was collected by Alex Krizhevsky et al. from Toronto.

Data Preparation

You load the training/testing datasets with the following command:

```
(x_train, y_train), (x_test, y_test) =
                        tf.keras.datasets.cifar10.load_data()
```

I will use 5% of the training dataset for validation during training and keep the testing dataset for model evaluation.

```
x_train, x_val, y_train, y_val = train_test_split(
                                    x_train, y_train,
                                    test_size = 0.05,
                                    random_state = 0)
```

In the situations, when the dataset does not contain enough images, we augment them by rotating, shifting, etc. to increase the count. The TensorFlow library provides a function for image augmentation; you just need to specify how you want to augment.

```
from tensorflow.keras.preprocessing.image
            import ImageDataGenerator
datagen = ImageDataGenerator(
 rotation_range = 15,
 width_shift_range = 0.1,
 height_shift_range = 0.1,
 horizontal_flip = True)
```

As all images are colored with color values ranging from 0 to 255, we will scale this data in the range of 0–1 with the following function:

```
def normalize(data):
    data = data.astype("float32")
    data = data/255.0
    return data
```

We augment and normalize our training set. We do not augment the validation and testing dataset, just normalize them.

```
datagen.fit(x_train)
x_train = normalize(x_train)
#datagen.fit(x_val)
x_val = normalize(x_val)
x_test = normalize(x_test)
```

We convert our target categorical column to numeric:

```
y_train = tf.keras.utils.to_categorical(y_train , 10)
y_test  = tf.keras.utils.to_categorical(y_test , 10)
y_val = tf.keras.utils.to_categorical(y_val , 10)
```

At this point, your data is reading for our experimentation. We will build the following networks:

- Our own CNN-based network
- Pre-trained VGG16
- Pre-trained Resnet50
- Pre-trained MobileNet
- Pre-trained DenseNet121

Modeling

We define few callbacks to control our trainings:

```
class myCallback(tf.keras.callbacks.Callback):
  def on_epoch_end(self, epoch, logs={}):
    if(logs.get('val_accuracy')>0.98):
      print(
          "\nReached 98% accuracy so cancelling training!")
      self.model.stop_training = True


callbacks = myCallback()
```

During training, if the validation accuracy reaches a level above 98%, we will stop further training.

The following callback will adjust the learning rate on the fly:

```
from keras.callbacks import ReduceLROnPlateau
lr_reduction = ReduceLROnPlateau
                            (monitor='val_accuracy',
                            patience=1,
                            verbose=1,
                            factor=0.5,
                            min_lr=0.000001)
```

The following callback will do the early stopping if there is no substantial improvement in validation accuracy over the last three epochs.

```
from keras.callbacks import EarlyStopping
early_stopping = EarlyStopping(monitor='val_accuracy',
                            min_delta=0.005,
                            patience=3,
                            verbose=1,
                            mode='auto')
```

We also set the optimizer for our networks:

```
optimizer = Adam(learning_rate=0.001,
                beta_1=0.9,beta_2=0.999)
```

With this we are now all set for experimenting the different architectures.

CNN-Based Network

We define our first network with the following code:

```
model=Sequential()
model.add(Conv2D(64,(3,3),strides=1,padding='Same',
              activation='relu',input_shape=(
              x_train.shape[1],x_train.shape[2],3)))
model.add(MaxPool2D(2,2))
model.add(BatchNormalization())
model.add(Conv2D(128,(3,3), strides=1,padding= 'Same',
              activation='relu'))
model.add(MaxPool2D(2,2))
model.add(BatchNormalization())
model.add(Flatten())
model.add(Dense(512, activation = "relu"))
model.add(Dropout(0.1))
model.add(Dense(10, activation = "softmax"))

model.compile(optimizer = optimizer ,
              loss = "categorical_crossentropy",
              metrics=["accuracy"])
```

Figure 18.19 shows the model summary.

The network consists of two convolution layers, followed by a flattening layer and then a dropout. The final layer is softmax activation for ten classes.

We train the model for 20 epochs. Note that our callbacks will take care of early stopping.

```
epoch = 20
history = model.fit(datagen.flow(
                    x_train, y_train, batch_size = 32
                    epochs = epoch,
                    validation_data = (x_val, y_val),
                    verbose = 1, callbacks=[
                    callbacks, lr_reduction, early_stopping])
```

Each epoch in my run (on GPU) took about 44 s and reached an accuracy of 78% after 13 epochs. The training stopped at 14 epochs due to our callback. You may tweak the parameters to the callback should you wish to train for more epochs in the hope of improving the accuracy further.

Figure 18.20 shows the accuracy and loss plots.

Figure 18.21 shows the confusion matrix on the validation set.

```
Model: "sequential"
_____
 Layer (type)                Output Shape              Param #
=================================================================
 conv2d (Conv2D)             (None, 32, 32, 64)        1792

 max_pooling2d (MaxPooling2D  (None, 16, 16, 64)       0
 )

 batch_normalization (BatchN  (None, 16, 16, 64)       256
 ormalization)

 conv2d_1 (Conv2D)           (None, 16, 16, 128)       73856

 max_pooling2d_1 (MaxPooling  (None, 8, 8, 128)        0
 2D)

 batch_normalization_1 (Batc  (None, 8, 8, 128)        512
 hNormalization)

 flatten (Flatten)           (None, 8192)              0

 dense (Dense)               (None, 512)               4194816

 dropout (Dropout)           (None, 512)               0

 dense_1 (Dense)             (None, 10)                5130

=================================================================
Total params: 4,276,362
Trainable params: 4,275,978
Non-trainable params: 384
```

Fig. 18.19 Model summary

Finally, we will check the loss and the accuracy score on the test dataset:

```
acc = model.evaluate(x_test , y_test)
print("test set loss : " , acc[0])
print("test set accuracy :", acc[1]*100)
```

This is the output

```
test set loss : 0.7102197408676147
test set accuracy : 77.10000276565552
training set loss : 0.5201990008354187
training set accuracy : 82.04420804977417
```

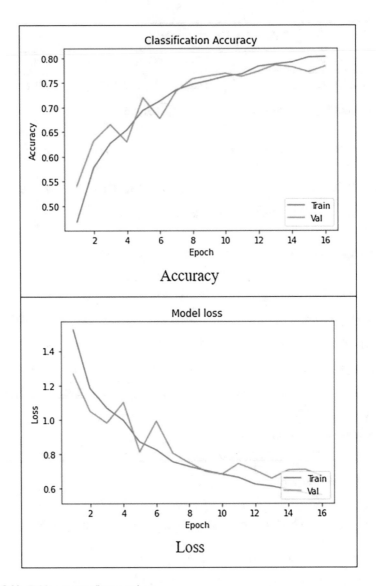

Fig. 18.20 Table accuracy/loss metrics

I tried the inference on a few unseen images. Figure 18.22 shows the results along with the confidence level of detection.

Next, we will try a pre-trained VGG16 model.

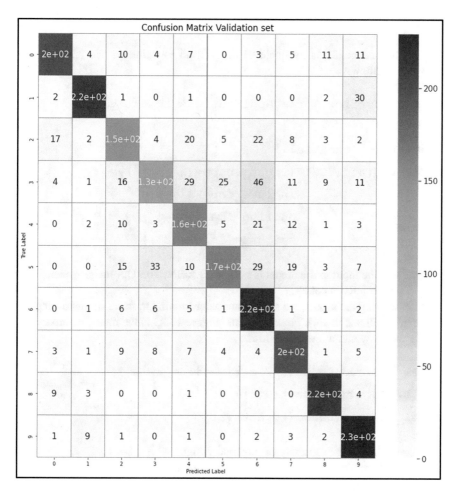

Fig. 18.21 Confusion matrix

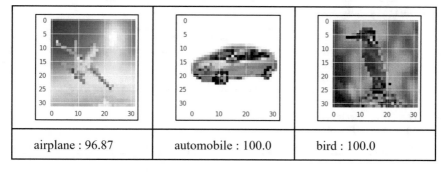

| airplane : 96.87 | automobile : 100.0 | bird : 100.0 |

Fig. 18.22 Inference on unseen images

VGG16

The pre-trained model is available in the TensorFlow library. We add it to our network as follows:

```
from tensorflow.keras.applications import VGG16
```

```
model_VGG=Sequential()
model_VGG.add(VGG16(input_shape=(32,32,3),
                    include_top=False,
                    pooling='max',
                    weights='imagenet'))
```

Note that the input to the model is a $32 \times 32 \times 3$ vector. Our image dimension is 32×32, with 3 RGB colors.

After loading the model, we set its parameter training to false, so that during our training, it will use its own preset weights. Figure 18.23 is the model summary.

We will now add a few layers to it.

```
model_VGG.add(Dense(512,activation='relu'))
model_VGG.add(Dropout(0.1))
model_VGG.add(Dense(10,activation='softmax'))
```

Figure 18.24 is the summary of our final design.
Note that we have only 267,786 trainable parameters.

```
Model: "sequential_1"
_____
Layer (type)                 Output Shape              Param #
=================================================================
 vgg16 (Functional)          (None, 512)               14714688

=================================================================
Total params: 14,714,688
Trainable params: 0
Non-trainable params: 14,714,688
_____
```

Fig. 18.23 Model summary

```
Model: "sequential_1"

 Layer (type)                  Output Shape              Param #
===============================================================
 vgg16 (Functional)            (None, 512)               14714688

 dense_2 (Dense)               (None, 512)               262656

 dropout_1 (Dropout)           (None, 512)               0

 dense_3 (Dense)               (None, 10)                5130

===============================================================
Total params: 14,982,474
Trainable params: 267,786
Non-trainable params: 14,714,688
```

Fig. 18.24 Model summary of the final design

We compile and run the model as in the earlier case. This is the accuracy report.

```
test set loss : 1.2783182859420776
test set accuracy : 55.640000104904175
training set loss : 0.5201990008354187
training set accuracy : 82.04420804977417
```

We now build our next model based on ResNet50.

ResNet50

Once again this pre-trained model is provided in TensorFlow Keras library. We add it to our network using the following code:

```
from tensorflow.keras.applications import ResNet50
model_RN=Sequential()
model_RN.add(ResNet50(input_shape=(32,32,3),
                      include_top=False,
                      weights='imagenet',
                      pooling='max'))
```

We add a few more layers to the network as in the earlier case. Figure 18.25 is the model summary.

```
Model: "sequential_2"

 Layer (type)                  Output Shape              Param #
========================================================================
 resnet50 (Functional)         (None, 2048)              23587712

 dense_4 (Dense)               (None, 512)               1049088

 dropout_2 (Dropout)           (None, 512)               0

 dense_5 (Dense)               (None, 10)                5130

========================================================================
Total params: 24,641,930
Trainable params: 1,054,218
Non-trainable params: 23,587,712
```

Fig. 18.25 Revised model summary

This is the accuracy score:

```
test set loss : 2.006408929824829
test set accuracy : 29.980000853538513
training set loss : 0.5201990008354187
training set accuracy : 82.04420804977417
```

You must be wondering why we are getting such a low accuracy in spite of using a pre-trained model. I will explain the results and provide you solutions for better accuracy after we run through all configurations. Next, we will try the MobileNet module.

MobileNet

We load the MobileNet module in our network using the following code:

```
from tensorflow.keras.applications import MobileNet
model_MN=Sequential()
model_MN.add(MobileNet(input_shape=(32,32,3),
                       include_top=False,
                       weights='imagenet',
                       pooling='max'))
```

```
Model: "sequential_3"

 Layer (type)                  Output Shape              Param #
=================================================================
 mobilenet_1.00_224 (Functio   (None, 1024)              3228864
 nal)

 dense_6 (Dense)               (None, 512)               524800

 dropout_3 (Dropout)           (None, 512)               0

 dense_7 (Dense)               (None, 10)                5130

=================================================================
Total params: 3,758,794
Trainable params: 529,930
Non-trainable params: 3,228,864
```

Fig. 18.26 Model summary of the new network

We add a few layers as before. Figure 18.26 is the model summary for our new network.

This is the accuracy score:

```
test set loss : 2.2873032093048096
test set accuracy : 13.70999962091446
training set loss : 0.5201990008354187
training set accuracy : 82.04420804977417
```

We now try one last configuration based on DenseNet121.

DenseNet121

We load and add the module to our network:

```
from tensorflow.keras.applications import DenseNet121
model_DN=Sequential()
model_DN.add(DenseNet121(input_shape=(32,32,3),
            include_top=False,
            weights='imagenet',
            pooling='max'))
```

```
Model: "sequential_4"

 Layer (type)                    Output Shape              Param #
=================================================================
 densenet121 (Functional)        (None, 1024)              7037504

 dense_8 (Dense)                 (None, 512)               524800

 dropout_4 (Dropout)             (None, 512)               0

 dense_9 (Dense)                 (None, 10)                5130

=================================================================
Total params: 7,567,434
Trainable params: 529,930
Non-trainable params: 7,037,504
```

Fig. 18.27 Model summary

As before, we add a few layers. Figure 18.27 is the summary of the final network. This is the accuracy report:

```
test set loss : 1.4147735834121704
test set accuracy : 51.010000705718994
training set loss : 0.5201990008354187
training set accuracy : 82.04420804977417
```

I could not include the Xception and InceptionV3 models in the above testing as Xception requires images of minimum size 71×71 and InceptionV3 requires 75×75 dimension images. I included these models in my further tests taken on high-resolution images, which I present next.

The final source for this project is available in the book's repository.

Summarizing Observations

I have summarized the observations on all the above models, which are presented in Table 18.1.

Looking at the accuracy values, you may wonder what is the use of using pre-trained models when the accuracies are so low. The reason that we have observed such low accuracies is because of the dataset we have used here. The CIFAR dataset, which we have used here, contains images of 32×32. All the pre-trained models were trained on images with much larger dimensions. Such datasets are not easily available in the public domain. Also, because of their large sizes, they require heavier resources and take longer time for vectorizing the images. Second, I had also put in some early stopping conditions. Adjusting these conditions and training for more epochs would have generated better results.

Table 18.1 Summarized test report of five pre-trained models

	CNN	VGG16	RecNet50	Mobile net	DenseNet121
Early stopping	13	18	NA	18	NA
Training time/epoch	~45 s	~50 s	~70 s	~43 s	~82 s
Training accuracy	78.26%	50.04%	19.20%	16.20%	46.29%
Validation accuracy	77.52%	52.22%	27.36%	17.40%	51.52%

Table 18.2 Testing report on high-quality images

	CNN by scratch	Xception	VGG16	ResNet50	InceptionV3	MobileNet
Epoch stopped	11	5	8	4	8	4
Training time	19 min, 28 s	9 min, 46 s	14 min, 43 s	8 min	14 min, 33 s	6 min, 15 s
Training accuracy	80.9%	92.5%	87.9%	64.4%	90.9%	93.9%
Validation accuracy	85.4%	94.6%	89.9%	67.3%	92.6%	94.4%

Modeling on High-Resolution Images

Just to show that the pre-trained models do produce good results, I ran the entire above experimentation on the Cats/Dogs dataset created by Microsoft, which is not in public domain. Table 18.2 presents the results.

With all said, using pre-trained models for image applications would certainly save you lots of time. You just have to use the appropriate image datasets.

Just to substantiate that the pre-trained models are indeed useful to data scientists, I developed another small project based on VGG16 to interpret a few images taken from the web.

Inferring Web Images

I used the Dog Breeds dataset created by Stanford. It classifies the dog into 120 different breeds. We will use the VGG16 pre-trained model to detect a dog and then extend it further to detect its breed. I used a few random images picked up from the web to check upon the classification.

The model was created by adding just one Dense layer of 512 dimensions with dropout to the pre-trained layer. The output layer as usual is Softmax. Figure 18.28 shows the model summary.

```
Model: "sequential"
_____
 Layer (type)                  Output Shape             Param #
================================================================
 vgg16 (Functional)            (None, 512)              14714688

 dense (Dense)                 (None, 512)              262656

 dropout (Dropout)             (None, 512)              0

 dense_1 (Dense)               (None, 5)                2565

================================================================
Total params: 14,979,909
Trainable params: 265,221
Non-trainable params: 14,714,688
```

Fig. 18.28 Model summary

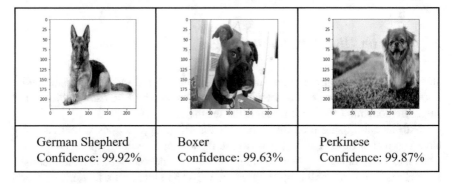

| German Shepherd
Confidence: 99.92% | Boxer
Confidence: 99.63% | Perkinese
Confidence: 99.87% |

Fig. 18.29 Inference on unseen images

It just took me less than 2 min to train the model on a reduced dataset comprising images of only five breeds of my interest. What I mean is that if your application requires identifying only German Shepherd, use only German Shepherd images for training. Why use images for all remaining 119 breeds. This reduced my training time substantially.

Now, for the testing, Fig. 18.29 presents the inference results along with their classification confidence for three random images picked up from the web.

Do you see with what accuracy the three dogs are identified? This is the power of using pre-trained models.

The entire project source is available in the book's repository.

Summary

We have several pre-trained models for both text and image datasets. It is beyond the limits of any individual data scientist to re-create such models. However, as the tech giants, who trained these models have made them available to the use of data scientists. Through some practical use cases, I showed how to use different pre-trained models in both categories. You observed that extending these models is easy, took just a few minutes to train them, and produced excellent accuracy on even totally unseen data. Go ahead and use pre-trained models for your own applications.

Chapter 19
Automated Tools

Data Scientist's Aid for Designing Classical and ANN-Based Models

Congratulations, by this time you have learned many techniques for developing AI models using both GOFAI and ANN. In this chapter, you will learn the use of a few tools that aid you in developing AI models.

In a Nutshell

In GOFAI, the selection of an appropriate algorithm and then fine-tuning its parameters has always been a daunting task for data scientists. In ANN, designing the optimal architecture has always been a challenge. Even the decision on when to select a pre-trained model versus designing your own network poses a big question. You will now learn tools that help you in model designs using both Classical AI and ANN.

Classical AI

There are plenty of classical algorithms available for training the machine. Even under the same category, like classification, clustering, and even regression, there are many algorithms. So far, the data scientists in the world were trying to get the best performing algorithm by either trial-n-error or their superb knowledge in this field. As the technology advanced, researchers created automated tools for testing and ranking algorithms based on their performance metrics. A data scientist can then pick up the best one or another high-ranking algorithm of his choice. Not only this, the tool also provides the ensemble of the top-ranking algorithms.

Every algorithm generally takes several parameters as its input. The algorithm performance varies largely based on the values of these parameters. These days, the automated tools would even fine-tune these hyper-parameter values by trying out

P. Sarang, *Thinking Data Science*, The Springer Series in Applied Machine Learning, https://doi.org/10.1007/978-3-031-02363-7_19

several combinations of values. The data-preprocessing and features engineering is also a vital step in the success of the AI model. These tools would also aid in designing the optimal data pipeline.

The algorithm selection, fine-tuning its hyper-parameters, and designing the data preprocessing pipeline are time-consuming processes. These tools take a few hours to a few days to execute, even on GPUs. However, it is worth the wait for a data scientist. Finally, the tool is simply an automated process that a data scientist previously used to do manually. Both commercial and free-to-use (open source) tools are available for designing models based on GOFAI.

I will now describe the use of an open-source tool called *auto-sklearn*.

Auto-sklearn

You want to develop a model for regression or classification, be it binary or multi-class, this open-source library will allow you to determine the best performing algorithm on your dataset. So, the target task can be any of:

- Binary classification
- Multiclass classification
- Multilabel classification
- Regression
- Multioutput regression

Using auto-sklearn is so easy that you just need to supply the training dataset to the model as shown here:

```
model = AutoSklearnClassifier(
                    time_left_for_this_task=4*60,
                    per_run_time_limit=30, n_jobs=-1)
model.fit(X_train, y_train)
print(model.sprint_statistics())
```

The first statement creates an instance of *AutoSklearnClassifier* that does everything that a data scientist would do in selecting the best performing algorithm. It does algorithm selection, hyper-parameter tuning, Bayesian optimization, meta-learning, and even ensemble construction. I will explain all these features in more depth shortly. The class takes a few parameters to control the execution. The *time_left +for_this_task* defines the time for each training task, which in this case is set to 4 min. The *per_run_time_limit* parameter sets the total runtime for the entire process. It is set to 30 min in the above example, beyond which further model development would stop.

Auto-sklearn for Classification on Synthetic Dataset

After we construct the model, we pass the training data to its *fit* method. After a substantially long time, yes, it can take even several hours for large datasets; it comes with the best performing model. You can verify this by printing the model's statistics. The run on a synthetic dataset that I created gives me the following output:

```
auto-sklearn results:
Dataset name: 744ab7e94247262949f04f5d879fc899
Metric: accuracy
Best validation score: 0.855856
Number of target algorithm runs: 55
Number of successful target algorithm runs: 55
Number of crashed target algorithm runs: 0
Number of target algorithms that exceeded the time limit: 0
Number of target algorithms that exceeded the memory limit: 0

Accuracy: 0.891
```

The dataset was synthesized for binary classification, and the created ensemble model has given an accuracy of 89%, fair enough in most of the cases. Consider that you have achieved all this, that is, building a machine learning model in just two lines of code. From the output, you can see that 55 algorithms were successfully applied on your dataset to rank them based on their performance and then ensembling them into the last model. Just imagine the time and efforts that you, as a data scientist, would have taken for this kind of experiment.

I tried the same thing on a multi-class dataset, a dataset having ten features. The accuracy was just 48%. The technology is still in development. And I am sure sklearn in the future will produce better models even for multi-class classification problems.

The entire source that produced the above-said results is available in the book's download repository.

I will now show you the results of the experiments that I performed on a real dataset, especially created for a binary classification problem.

Auto-sklearn for Classification on Real Dataset

To demonstrate the AutoML for a standard binary classification task, I used the gesture dataset deployed on Kaggle. The dataset was designed to map prosthetic gestures such as open/close hand or rotate wrist. Initially, I tried the *SVC* (support vector classifier) algorithm on this dataset; it gave me an accuracy of 97.8%. Then, I tried the *AutoSklearnClassifier* on it. The output is shown here:

```
auto-sklearn results:
Dataset name: 5845bd51833f800db6791112c87423a6
Metric: accuracy
Best validation score: 0.990256
Number of target algorithm runs: 25
Number of successful target algorithm runs: 17
Number of crashed target algorithm runs: 0
Number of target algorithms that exceeded the time limit: 8
Number of target algorithms that exceeded the memory limit: 0
```

Accuracy: 0.993

As you see, it used 25 algorithms for training. The last model gave me an accuracy of 99%. A wonderful improvement. Going from 97% to 99% is lot more difficult than improving accuracy from 50% to 70%. You can ask auto-sklearn tool which model did it select as a final one by calling its *show_models* method.

```
model_auto.show_models()
```

This is the partial output produced by the above statement:

```
'[(0.200000, SimpleClassificationPipeline({'balancing:strategy':
'none', 'classifier:__choice__': 'extra_trees', 'data_preprocessing:
categorical_transformer:categorical_encoding:__choice__':
'one_hot_encoding', 'data_preprocessing:categorical_transformer:
category_coalescence:__choice__': 'no_coalescense',
'data_preprocessing:numerical_transformer:imputation:strategy':
'median', 'data_preprocessing:numerical_transformer:rescaling:
__choice__': 'minmax', 'feature_preprocessor:__choice__':
'no_preprocessing', 'classifier:extra_trees:bootstrap': 'True',
'classifier:extra_trees:criterion': 'gini', 'classifier:extra_trees:
max_depth': 'None', 'classifier:extra_trees:max_features':
0.5033866291997137, 'classifier:extra_trees:max_leaf_nodes': 'None',
'classifier:extra_trees:min_impurity_decrease': 0.0, 'classifier:
extra_trees:min_samples_leaf': 2, 'classifier:extra_trees:
min_samples_split': 14, 'classifier:extra_trees:
min_weight_fraction_leaf': 0.0},\ndataset_properties={\n 'task': 1,
\n 'sparse
```

If you study the output closely, you notice that it has used an *ExtraTrees* classifier. The best values for all hyper-parameters are listed too. I even tried comparing the classification report for both approaches—the *SVC* and the auto-sklearn. Figure 19.1 show the comparison of two classification reports.

The entire source that produced the above-said results is available in the book's download repository.

Next, I will show you how to use auto-sklearn for regression tasks.

	precision	recall	f1-score	support
0	0.98	0.97	0.98	440
2	0.98	0.98	0.98	438
accuracy			0.98	878
macro avg	0.98	0.98	0.98	878
weighted avg	0.98	0.98	0.98	878

SVC classification report

	precision	recall	f1-score	support
0	1.00	0.99	0.99	419
2	0.99	1.00	0.99	459
accuracy			0.99	878
macro avg	0.99	0.99	0.99	878
weighted avg	0.99	0.99	0.99	878

Auto-sklearn classification report

Fig. 19.1 Comparison of classification reports by SVC and auto-sklearn

Auto-sklearn for Regression

For this task, I used the insurance dataset deployed on Kaggle. The dataset has six features and one target called charges. As in the case of classification task, I first tried the linear regression implementation of sklearn with the following code:

```
from sklearn.linear_model import LinearRegression

linearRegression = LinearRegression().fit(
                        aX_train_scaled, label_train)

y_pred = linearRegression.predict(X_val_scaled)

error_metrics(y_pred,label_val)
```

It produced the following output:

```
MSE: 0.1725732138508936
RMSE: 0.4154193229146829
Coefficient of determination: 0.7672600627397954
```

Then, I tried auto-sklearn with the following code:

```
model_auto_reg = AutoSklearnRegressor(
                        time_left_for_this_task=4*60,
                        per_run_time_limit=30,
                        n_jobs=-1)
model_auto_reg.fit(X_train_scaled,label_train)
print(model_auto_reg.sprint_statistics())


y_pred_auto = model_auto_reg.predict(X_val_scaled)
```

It gave me the following output:

```
auto-sklearn results:
 Dataset name: 18c15b94e55d0222e0ab55cfcf68e850
 Metric: r2
 Best validation score: 0.785246
 Number of target algorithm runs: 40
 Number of successful target algorithm runs: 36
 Number of crashed target algorithm runs: 0
 Number of target algorithms that exceeded the time limit: 4
 Number of target algorithms that exceeded the memory limit: 0
```

As you see, it tested 40 algorithms out of which 36 ran successfully. The best validation score was 78.5%, and the metric used for measurement was R-2 score. I called *error_metrics* to check the performance.

```
error_metrics(y_pred_auto,label_val)
```

This is the output:

```
MSE: 0.11222892540046188
RMSE: 0.33500585875542815
Coefficient of determination: 0.8495015171557208
```

Figure 19.2 gives an immediate comparison between the two approaches.

Using coefficient of determination as a reference, you can easily see that the auto-sklearn outperformed the manual model based on linear regression. Now, let us check out which model the auto-sklearn finally selected.

```
MSE:   0.1725732138508936
RMSE:  0.4154193229146829
Coefficient of determination:  0.7672600627397954
```

Error metrics for LinearRegression

```
MSE:   0.11222892540046188
RMSE:  0.33500585875542815
Coefficient of determination:  0.8495015171557208
```

Error metrics for auto-sklearn

Fig. 19.2 Comparison of error metrics for LinearRegression and auto-sklearn

```
model_auto_reg.show_models()
```

This is the output:

```
'[(0.340000, SimpleRegressionPipeline({'data_preprocessing:
categorical_transformer:categorical_encoding:__choice__':
'no_encoding', 'data_preprocessing:categorical_transformer:
category_coalescence:__choice__': 'minority_coalescer',
'data_preprocessing:numerical_transformer:imputation:strategy':
'median', 'data_preprocessing:numerical_transformer:rescaling:
__choice__': 'quantile_transformer', 'feature_preprocessor:
__choice__': 'feature_agglomeration', 'regressor:__choice__':
'random_forest', 'data_preprocessing:categorical_transformer:
category_coalescence:minority_coalescer:minimum_fraction':
0.458966132642697, 'data_preprocessing:numerical_transformer:
rescaling:quantile_transformer:n_quantiles':
410, 'data_preprocessing:numerical_transformer:rescaling:
quantile_transformer:output_distribution': 'normal',
'feature_preprocessor:feature_agglomeration:affinity': 'cosine',
'feature_preprocessor:feature_agglomeration:linkage': 'average',
'feature_preprocessor:feature_agglomeration:n_clusters':...'
```

As you observe, it selected a *random_forest* regressor. The output also shows the best values for its hyper-parameters. I also created a regression plot based on the auto-sklearn generated model. The output is shown in Fig. 19.3.

As you see from both above experiments that the AutoML technology has eased the data scientist's task of selecting the best algorithm to be used on a specified dataset and developing a best performing machine learning model.

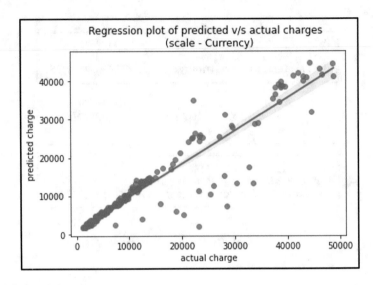

Fig. 19.3 Regression plot

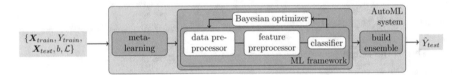

Fig. 19.4 Auto-sklearn architecture. (Image source: https://github.com/automl/auto-sklearn)

The entire source that produced the above-said results is available in the book's download repository.

I will now discuss the auto-sklearn architecture.

Auto-sklearn Architecture

Auto-sklearn is a library for AutoML that uses models implemented in sklearn machine learning library. It does both algorithm selection and hyper-parameter optimization. It also has a meta-learning step at the beginning and an automated ensemble construction at the end. We depict this in Fig. 19.4.

The meta-learning step decides on the best performing data pipeline. They based it on the idea of collaborative filtering and Bayesian optimization. The collaborative filtering is based on the idea that if two persons A and B like the same item, it is more likely that A would like another item that B likes. The collaborative filtering encompasses techniques for matching data pipelines with datasets. Nicolo Fusi et al. experimented this kind of matching on 5000 ML pipelines across 576 OpenML datasets. Thus, given your dataset, the meta-learning step automatically decides upon the optimal pipeline.

At the end, the system constructs an ensemble of all the models it tried out. The idea is to save all the hard-work done on training those models. Rather than selecting the top-ranking model as the final one, an ensemble of high-ranking models is constructed.

I will now describe some of the salient features of auto-sklearn.

Auto-sklearn Features

Here are some of the salient features of auto-sklearn:

- Allows you to include/exclude specific algorithms in the auto-test.
- Allows you to set the number of algorithms in the test.
- As the entire process takes a long time to run, you can set your preferred process time beyond which the testing would stop.
- Certain algorithms may take a long time for their training. You can set the time limit to train a model.
- Allows you to set the size of an ensemble model.
- Allows you to specify a sampling strategy—split, cross-validation, partial CV.
- Allows parallelization to run the tests concurrently.

Though I have taken the sklearn implementation to show you the power of AutoML, there are many other tools available in the market—both commercial and free-to-use open source. I will describe their features also toward the end that will help you in selecting a tool of your choice.

What's Next?

The auto-sklearn designs a best performing machine learning model based on classical algorithms. What about the ANN/DNN approach for model building? Designing ANN architecture, as you have seen earlier, is not a simple task. Fortunately, we now have a technology for auto-design of an ANN model. The AutoKeras is such an AutoML system based on Keras, which I will describe next.

ANN/DNN

We have seen that the auto-sklearn selects the best performing algorithm among a set of known algorithms. In case of neural networks, the network architecture is the key portion of the model development. One has to try out several architectures to decide on the best performing one. The AutoKeras does this for you as an automated process. It evaluates a set of different architectures against your dataset and, at the end, gives you the best performing one. So, even if you do not have the knowledge of machine learning, with the help of AutoKeras, you will create machine learning models based on ANN technology.

Like auto-sklearn, AutoKeras allows you to perform many kinds of tasks. These are listed here:

- Image classification
- Image regression
- Structured data classification
- Structured data regression
- Text classification
- Text regression
- Time series forecasting

I will show you how to use AutoKeras system on both regression and classification tasks.

AutoKeras for Classification

Before you experiment on AutoKeras, you need to install couple of packages as shown here:

```
!sudo pip install git+https://github.com/keras-
team/keras-tuner.git@1.0.2rc1
!pip install autokeras
```

After a successful installation, you are ready to create a neural network architecture for a classification task. I will use the fetal health dataset taken from Kaggle. The dataset has around 2000 rows and 22 columns. The last one called *fetal_health* is our target, which has three different values—three classes for our model. As the dataset is not balanced, I used SMOTE for balancing it. You can refer to the full source in the download section. After creating the training and testing datasets and applying the standard scalar, I applied the classifier on it. This is the code for using the classifier.

```
from autokeras import StructuredDataClassifier
```

```
search = StructuredDataClassifier(max_trials=10,
                                  num_classes=3)
search.fit(x=X_train_s, y=y_train, verbose=0, epochs=5)
```

The *StructuredDataClassifier* designs the network for you. You need to specify the number of classes you want the classifier to output.

After it finds the best architecture, you can evaluate the model's accuracy.

```
loss, acc = search.evaluate(X_test_s, y_test,
                            verbose=0)
print('Accuracy: %.3f' % acc)
```

This is the output:

```
Accuracy: 0.944
```

Good enough, you may increase the number of trials to see if the accuracy improves further. Use the model for inference on the test data and print the classification report.

```
y_predicted = search.predict(X_test_s)
print(classification_report(y_test,y_predicted))
```

Figure 19.5 shows the classification report.

You can now check the created network architecture by printing its summary report.

	precision	recall	f1-score	support
0	0.95	0.93	0.94	520
1	0.92	0.92	0.92	485
2	0.96	0.99	0.98	485
micro avg	0.94	0.94	0.94	1490
macro avg	0.94	0.94	0.94	1490
weighted avg	0.94	0.94	0.94	1490
samples avg	0.94	0.94	0.94	1490

Fig. 19.5 Classification report

```
Model: "model"

Layer (type)                    Output Shape              Param #
=================================================================
input_1 (InputLayer)            [(None, 252)]             0

multi_category_encoding (Mul    (None, 252)               0

normalization (Normalization    (None, 252)               505

dense (Dense)                   (None, 32)                8096

re_lu (ReLU)                    (None, 32)                0

dense_1 (Dense)                 (None, 256)               8448

re_lu_1 (ReLU)                  (None, 256)               0

dense_2 (Dense)                 (None, 3)                 771

classification_head_1 (Softm    (None, 3)                 0
=================================================================
Total params: 17,820
Trainable params: 17,315
Non-trainable params: 505
```

Fig. 19.6 Model summary

```
model = search.export_model()
model.summary()
```

Figure 19.6 shows the model summary.
You can also plot the network model using following statement:

```
plot_model(model, show_shapes=True,
           show_layer_names=True)
```

Figure 19.7 shows the network model.

As you see, it has created a three-dense-layer network with of course an input layer and a softmax for the output. The activation used on all three dense layers is ReLU. The first two dense layers consist of 32 and 64 nodes.

You will appreciate how neatly it has designed the network on its own, saving you the trouble of experimenting on different architectures.

The entire source that produced the above-said results is available in the book's download repository.

Let us now design a network using AutoKeras for a regression task.

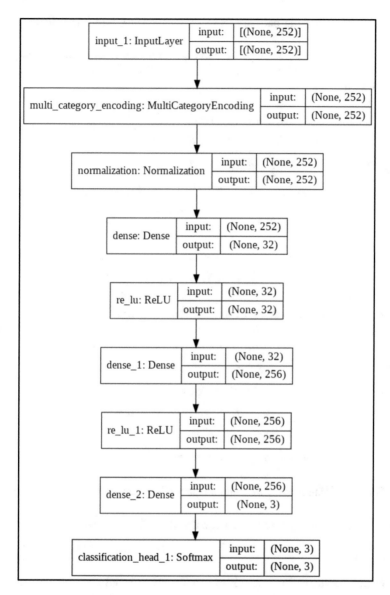

Fig. 19.7 Network architecture generated by AutoKeras for classifier

AutoKeras for Regression

For regression, I have used the insurance cost dataset from Kaggle. It has 7 columns, the charges column is our target. The target value depends on the 6 remaining columns, which are our features. We need to find a regression hyperplane to establish the relationship between these six features and the insurance charges, our target.

For regression tasks, we use *StructuredDataRegressor*, which takes two parameters—the number of trials and the loss function. We define a callback for our loss function as follows:

```
from tensorflow.keras.callbacks
            import ReduceLROnPlateau

lr_reduction = ReduceLROnPlateau(
                monitor='mean_squared_error',
                patience=1,
                verbose=1,
                factor=0.5,
                min_lr=0.000001)
```

We now invoke the regressor using the following code:

```
regressor = StructuredDataRegressor(
            max_trials=3,
            loss='mean_absolute_error')
regressor.fit(x=X_train_scaled, y=label_train,
            callbacks=[lr_reduction],
            verbose=0, epochs=200)
```

After the training is over, we evaluate the model's performance on the test dataset.

```
mae, mse = regressor.evaluate(X_val_scaled,
                    label_val, verbose=0)
```

You can do the performance testing on the test data and print the error metrics.

```
label_predicted = regressor.predict(X_val_scaled)
error_metrics(label_predicted,label_val)
```

This is the output:

```
MSE: 0.1559527860627597
RMSE: 0.3949085793734541
Coefficient of determination: 0.820735892953915
```

```
Model: "model"

Layer (type)                    Output Shape           Param #
=================================================================
input_1 (InputLayer)            [(None, 14)]           0

multi_category_encoding (Mul    (None, 14)             0

normalization (Normalization    (None, 14)             29

dense (Dense)                   (None, 32)             480

re_lu (ReLU)                    (None, 32)             0

dense_1 (Dense)                 (None, 32)             1056

re_lu_1 (ReLU)                  (None, 32)             0

regression_head_1 (Dense)       (None, 1)              33
=================================================================
Total params: 1,598
Trainable params: 1,569
Non-trainable params: 29
```

Fig. 19.8 Model summary

Now, we will check out the network designed by AutoKeras.

```
model = regressor.export_model()
model.summary()
```

Figure 19.8 shows the model summary.
You can also plot the model.

```
plot_model(model, show_shapes=True,
            show_layer_names=True)
```

Figure 19.9 shows the network architecture.

The entire source that produced the above-said results is available in the book's download repository.

So far, you studied the classifier and regressor on numeric structured data. I will now show you how to create a classifier on image data.

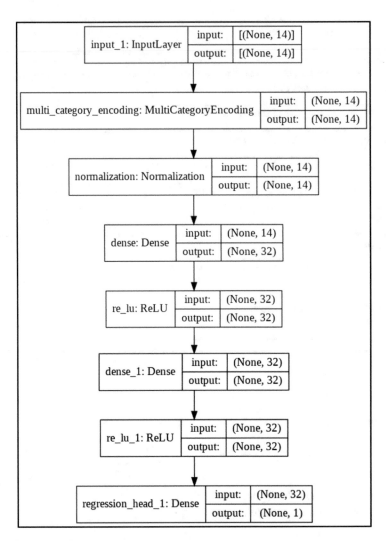

Fig. 19.9 Network architecture generated by AutoKeras for regression

AutoKeras Image Classifier

For demonstration of *ImageClassifier*, I will use the famous MNIST dataset containing images of handwritten digits 0 through 9. As earlier, you specify the number of trials and the number of classes in *ImageClassifier* construction. This is the code:

```
import autokeras as ak
from autokeras import ImageClassifier

clf = ak.ImageClassifier(num_classes=10,
                         overwrite=True,
                         max_trials=1)
history=clf.fit(
    x_train,
    y_train,
    validation_split=0.15, epochs=10)
```

The *overwrite* parameter, when set to false, reloads an existing project of the same name if one is found. After the model training, you can evaluate its performance on the test data.

```
predicted_y = clf.predict(x_test)
```

We can print the loss and accuracy scores on the ten classes:

```
pd.DataFrame(history.history)
```

Figure 19.10 shows the output.

Fig. 19.10 Loss and accuracy for the ten classes

	loss	accuracy
0	0.155441	0.952450
1	0.071200	0.977583
2	0.057938	0.982267
3	0.051609	0.983817
4	0.045898	0.985300
5	0.039828	0.987067
6	0.036760	0.988850
7	0.035004	0.988900
8	0.032611	0.989617
9	0.030548	0.989600

The plot for model's performance on various epoch is shown in Fig. 19.11. You may also print the classification report as follows:

```
print(classification_report(y_test,
                predicted_y.astype(np.uint8)))
```

Figure 19.12 shows the classification report.

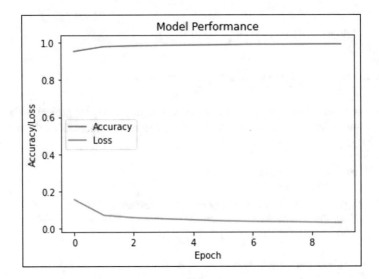

Fig. 19.11 Performance metrics for AutoKeras image classifier

	precision	recall	f1-score	support
0	0.99	0.99	0.99	980
1	0.99	1.00	0.99	1135
2	0.98	0.99	0.99	1032
3	0.99	0.99	0.99	1010
4	0.99	0.99	0.99	982
5	0.98	0.99	0.99	892
6	1.00	0.98	0.99	958
7	0.99	0.98	0.99	1028
8	0.98	0.99	0.99	974
9	0.99	0.98	0.99	1009
accuracy			0.99	10000
macro avg	0.99	0.99	0.99	10000
weighted avg	0.99	0.99	0.99	10000

Fig. 19.12 Classification report

The entire source that produced the above-said results is available in the book's download repository.

You have so far seen the use of Auto-sklearn and AutoKeras. I will now list a few more AutoML tools that offer varying capabilities.

More AutoML Frameworks

The auto-sklearn with minimal input parameters is one of the simplest to use. It is open source and implemented on top of the popular library sklearn. It designs the ML models based on classical algorithms.

AutoKeras is again an open-source library based on top of sklearn that designs neural network architectures for developing ML models. This too is easy to use.

PyCaret

PyCaret is an open source tool that is based on the R library. It provides data preparation, features engineering, features selection, modeling, analysis, and deployment. This too requires minimum inputs from you and thus is fairly automated. It can handle structured supervised and unsupervised tasks. It supports 43 algorithms (as of this writing) that cover both classifiers and regressors. Most of the algorithms that you have studied in this book so far are included in their testing, except for the fact that the implementation is R-based. The best part is that it allows the model comparisons based on their error metrics. It allows the analysis and deployment of a chosen model with a graphical interface and less coding.

MLBox

This too is an open-source library based on Python. It is comparatively fast in training models. It uses seven strategies—LightGBM, random forest, ExtraTrees, Tree, bagging, AdaBoost, and linear (classifier and regression). It does basic data preprocessing and model optimizations and handles structured supervised tasks. It probably requires more coding. Hopefully, in the near future, we will see more enhancements to this library by the community contributors.

TPOT

This is an open source Python library, also known as data science assistant. This tool explores thousands of data pipelines to select the most appropriate one for your dataset. It includes features engineering, data preprocessing, modeling, and optimizations. They mainly aimed it at structured supervised and image classification tasks. The function requires you to input only three arguments and find the best pipeline for the error metrics specified by you. As of this writing, it is still under development and widely supported by the data science community.

H2O.ai

The frameworks that you have seen so far are Python or R-based. H2O is one that supports R, Python, Java, and Scala, so you have a wider choice of programming languages. Not only, this also provides a larger set of automations. It allows complex data processing, model development, model evaluation, ensembling, web, and mobile application development and even has a low code framework. Some of these services are paid one.

The model development is mostly GUI based, requiring minimal coding on our side. It supports structured and unstructured datasets and both supervised and unsupervised tasks. It also allows model comparisons with the help of appealing graphics. Most of the time you will use its GUI interface, and rarely you would be required to do some hand-coding. A wide community, which has also published several use cases and tutorials, supports the project.

DataRobot

This is a cloud platform that offers many paid services. The services include data engineering, machine learning, MLOps, augmented intelligence, and others. For data science projects, it allows the full development of the data pipeline and the deployment of ML models on the cloud. It supports structured and unstructured datasets and both supervised and unsupervised tasks. Its user-friendly GUI allows easy configuration of all processes and creates an explainable project.

DataBricks

Like DataRobot it offers a wide variety of paid services. For AutoML it supports data preprocessing, features engineering, hyper-parameter tuning, and best model selections. It can build models based on both classical ML and ANN/DNN. It allows coding using PySpark. It supports structured and unstructured datasets and both supervised and unsupervised learning. Particularly, it can work on big data. This too has a very user-friendly GUI for process configurations and creating explainable projects.

BlobCity AutoAI

This is the latest one I came across at the time of writing this chapter. It is really simple to use. You just upload your dataset and ask it to do a model building on it. It, on its own, experiments both types of model development—classical ML and ANN. It decides between regression and classification. It provides ensembling and hyperparameter tuning. The best part of it, that I did not see in any of the other tools that I tried so far, is spilling out the entire source. I consider it an enormous advantage for data scientists. Data scientists can fiddle with the code and, most importantly, submit it to their clients, claiming it to be their own creation. As of this writing, the project was launched just a few weeks ago. I would expect lots of improvements, features additions, and use cases, by the time you read this.

Summary

In this chapter, you studied an automated process that is important for data scientists in quickly narrowing down on their search for the best machine learning model. AutoML technology provides this automation. It supports model building both in classical ML and ANN. You studied the various tools available in the market, focusing on auto-sklearn and AutoKeras to explain the technology. Do your own evaluations of the tools and use them practically to ease your model development. A bit of warning again, the technology is still under development, so be sure to look at the tool vendor's latest updates.

Chapter 20
Data Scientist's Ultimate Workflow

A Quick Summary on a Data Scientist's
Approach to Model Development

The first chapter started with the data science process, where I defined the various stages of developing ML applications. The following chapters covered the details at each stage. You learned data analytics, data cleansing, several dimensionality reduction techniques, data preparation, and a wide range of classical ML algorithms covering both regression and classification tasks, followed by an exhaustive coverage on the clustering algorithms.

The ANN technology brought in a paradigm shift to classical statistical ML (GOFAI) technology. The data scientist now has a choice between GOFAI and ANN for model development. In ANN itself, we observed two paths—constructing your own networks or to go in for pre-trained models, with a technology that we call as transfer learning. The whole AI development field, as you have learned, is quite wide, and it could be really hard for a data scientist to decide on the most rewarding path in the development.

I will now summarize all my teachings and provide you with definite guidelines that would help you develop the most efficient machine learning models of your dreams.

Consolidated Overview

I present here the consolidated workflow as a cheat sheet, so that you know at a glance what would be your action path while developing AI applications. The consolidated workflow is presented in Fig. 20.1.

You had seen these workflows in Chap. 1. The above schematic provides the consolidation and the links between the workflows. I will present you with definite guidelines to use this cheat sheet. Before I begin with the workflow-1 in the above sheet, I am going to introduce you to another important decision-making workflow that I have termed workflow-0.

© The Author(s), under exclusive license to Springer Nature Switzerland AG 2023   351
P. Sarang, *Thinking Data Science*, The Springer Series in Applied Machine Learning,
https://doi.org/10.1007/978-3-031-02363-7_20

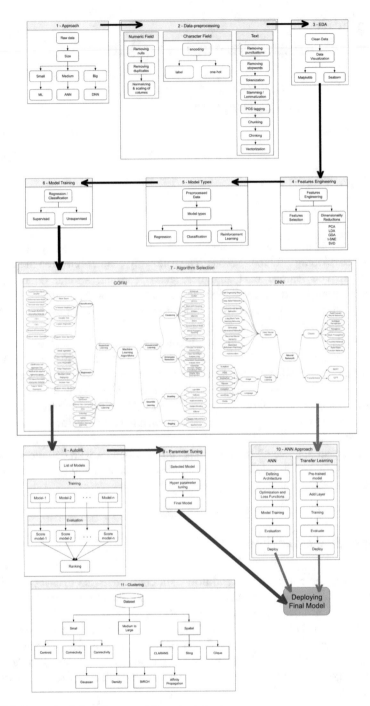

Fig. 20.1 Data scientist's ultimate workflow cheat sheet

Workflow-0: Quick Solution

I have depicted in the new workflow in Fig. 20.2.

As a data scientist, when I get the data from the client and after listening to his business requirements, the first thing I try is to use automated tools. I mentioned MLaaS in Chap. 1; I use this service to create the so-called "quick-n-dirty" solution. Though this is a general term, my observations are that the solutions given by these services are not dirty. These services have done a good job for me. The only issue is that I cannot sell their solution *as is* to my customer for a fee. If the customer does not have any issue in paying for your quick solution, go ahead and deploy the suggested model on their servers for production use. Another important aspect here is that the customer should be willing to upload his full dataset on their servers.

If I cannot overcome the above restrictions, I go for an automated tool; I had discussed several such tools in Chap. 19. You just ingest the data into their tool and they will spill out the best performing model. As I mentioned during the discussion on these tools, such tools do everything from creating efficient data pipelines to ensembling an ultimate model. Some of the more advanced tools even decide between GOFAI and ANN technologies. As an ordinary user of a tool, I do not even have to clean my data. Now, again, the question remains, can I sell my model to the client without having its source?

Fig. 20.2 Workflow-0: quick solution

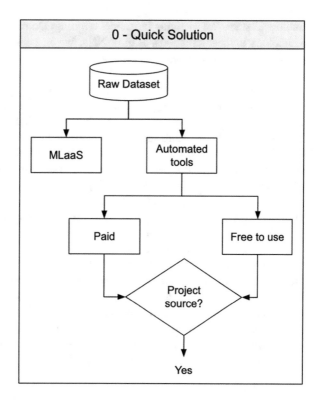

As I mentioned earlier, I have evaluated BlobCity AutoAI that spills out the full project source for the generated model. What more anybody would ask for? I used this source as my original creation and charge money to the customer, without him knowing how I did it. I am wondering whether, at the time of you reading this, will there be more tools competing with BlobCity? By the way, BlobCity is open-sourced, and I sincerely hope that the community helps them in keeping the product alive for years to come.

I wonder, if the business owners come to know of these tools, we, the data scientist, would be out of a job. But that will not happen as though these services can easily be used by a layperson (nontechnical persons), businesses are always sensitive to data sharing. And they also look at the service support for many years, with periodic updates to their applications (models) to account for dynamic nature of data over many years.

Concluding that we, data scientists, are still in the game, I will now proceed with the first workflow, followed by a data scientist.

We begin with the workflow-1 marked in the above diagram.

Workflow-1: Technology Selection

First, ask your client how many data points he has for sharing with you. If this number is too small, comprising just a few hundreds to little over thousands, stick to traditional statistical methods for model building. Do not even think of using ANN technology in these situations, as the ANNs, as you know now, have many trainable parameters and you need sufficiently large datasets to fine-tune those hyper-parameters.

If you have a few thousands and tens of thousands of data points available, think of deploying ANN technology. Training GOFAI algorithms on such size datasets would take lots of resources and training time. Consider using ExtraTrees or XGBoost kind of boosting algorithms applied to such datasets; they do take a considerable amount of training time. In the end, they still produce excellent results. However, if ANN technology can give you similar results in less time, why not use it? So, first try out the ANN technology on medium-sized datasets.

Now, what if your client is an online merchandise seller, who can provide you millions of data points for analytics? Here, the only option that is left for you is to go in for DNN. You will construct your own architecture consisting mostly of many dense layers followed by dropouts, the last layer being softmax in many situations. The network would take a long time to train; but ultimately you do not have any other option. If you are lucky enough to find a pre-trained model in the area of interest, use it. This is a rare situation, unless your application needs is for text or image datasets, where we have lots of pre-trained models at our disposal. If your current need is to develop a sentiment analysis, document summarization, news classification, etc. use any of the pre-trained text-based models that I have discussed previously in the book. If your need is for an image-based application, like identifying dog breeds or face detection, use one of the several models that I discussed in earlier chapters.

After you have decided on the type of technology, GOFAI versus ANN, your next workflow is data cleansing.

Workflow-2: Data Preprocessing

The data cleansing process, as depicted in workflow-2, depends on the type of data.

For a numeric data field, you would first remove nulls and duplicates. You may impute missing values or eliminate those columns altogether, having many missing fields. After this, you will normalize the data to bring the range of all data columns to the same scale.

For character-based columns, you encode those into numeric format by using the label or one-hot encoder.

If your application is based on a huge-text corpus, you will need to employ several NLP techniques to bring down the vocabulary size, which ultimately represents the features for model training.

After cleaning the data, you proceed to the next workflow—EDA.

Workflow-3: EDA

In this workflow, you do an exploratory data analysis to understand your data. Here, domain knowledge plays a vital role. I usually call for a domain expert from the customer's side to understand the data and have a better understanding of the customer's business needs. I mostly use Matplotlib and sometimes Seaborn for more advanced charts. There are many more advanced tools available for data visualization. In the dimensionality reduction chapter, I had discussed a few techniques for visualizing high-dimensional datasets.

After understanding the data, I move on to the important workflow, and that is features engineering.

Workflow-4: Features Engineering

Having more features at your disposal looks like an advantage; however, in machine learning, we call it "curse of dimensionality." In Chap. 2, I had discussed several methods of reducing dimensions in your dataset. We mainly reduce the dimensions for two purposes—visualizing data in 2-D or 3-D space and creating more efficient models with less over-fitting. I used both features selection and dimensionality reductions techniques discussed in the chapter.

My next task is selecting the model type.

Workflow-5: Type of Task

I look at the client requirement to decide on what type of machine learning model would meet the client's requirements. The choice is between regression and classification. The reinforcement requirement is rare and if you ever come across those kinds of requirements, you will need to learn the algorithms in this space. I have come across requirements for designing recommendation systems where the better choice was reinforcement learning.

After deciding on the model type, I move ahead to the next task.

Workflow-6: Preparing Datasets

My next task is to label the dataset. In some situations, the client may supply you with the labeled dataset. If not, I would do it at least for a subset. This applies only to regression and classification tasks. The clustering comes in a totally different category. Having or not having a labeled dataset is a deciding factor between supervised and unsupervised learning.

Now, I am fully ready for model development.

Workflow-7: Algorithm Selections

This is the major task in model development. Usually, I start out with a simple algorithm like random forest. If the results are good, I move forward to boosting algorithms. In many situations, ExtraTrees and XGBoost have given me excellent results. If I still need better accuracy, I would try other algorithms in this category and experiment in creating ensemble models. In case of regression, this kind of experimentation is not usually required. Regression is the most-studied field of statistics and implementers have done an excellent job in implementing those algorithms.

This workflow branches into two paths—classical ML development takes me to workflow-8 and ANN approach branches to workflow-10. I now discuss workflow-8.

Workflow-8: AutoML

In the model selection in workflow-7, the best approach that I find is to take help of automated tools in arriving at some quick decisions. Usually, these tools produce an ensemble model as a last model. I look up their suggestions and start building my own. While building it on my own, I need to follow workflow-9.

Workflow-9: Hyper-parameter Tuning

If you do not fine-tune the hyper-parameters, the algorithm will not produce excellent results. I have already discussed methods like *GridSearchCV* for fine-tuning the parameters. Sometimes, I use Optuna to get the optimal values of the parameters.

After lots of experimentation on ranking algorithms, fine-tuning their hyper-parameters, and ensembling, I conclude upon a last model, ready for production use. So, move it to the production server.

Now, if I have decided upon the ANN approach in workflow-7, I follow the path to workflow-10.

If I come across a clustering problem, I treat it totally differently by applying the various unsupervised clustering algorithms that I have discussed under workflow-11.

Workflow-10: ANN Model Building

If I am using the ANN technology, the call is between designing my network or to use a pre-trained model.

Given the liberty of having additional data points, I decide upon the ANN technology. Having millions of data points and gigabytes of data, one has to resort to deep neural networks, where the architectures are complex and require several weeks of training time. Fortunately, a data scientist can take advantage of the pre-trained models by deploying transfer learning techniques.

I have already discussed in a previous chapter how to design your own network and how to use a pre-trained model. I have also provided you a list of pre-trained models (the list is not exhaustive and is unbiased) for both text and image-based applications.

After training and evaluating the ANN/DNN model, if you reach an acceptable level of accuracy, just proceed to the last step of deploying the model to a production server.

There is one final workflow that I have not included in the above consolidation and that is clustering.

Workflow-11: Clustering

As said earlier, the notion of cluster is not well defined. There is no definite measure in cluster formations and especially there may not be a consensus amongst all the stake-holders. You would apply different clustering algorithms to the dataset, share your results with the client and provide him with different perspectives. Let him take his final decision and make his own conclusions. Whenever the client clearly defined

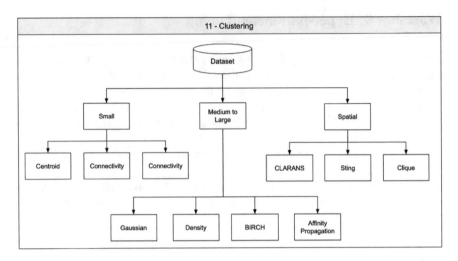

Fig. 20.3 Workflow-11: clustering.

the similarity measure for determining the clustering quality, I could provide results satisfying his business needs.

Depending on the size of your dataset, follow the path stated in the schematic presented in Fig. 20.3.

Clustering small- to medium-sized datasets is not too complex and the simple algorithms like K-means and K-medoids work well. With known Gaussian distributions, the GMM algorithm works well. For medium to large datasets, density-based algorithms and others like BIRCH and affinity propagation do a good job. For spatial, huge datasets, clustering is very complex, and the process is resource-hungry and time-consuming. Usually, you will divide the full data space into subspaces and then apply those advanced algorithms to cluster the full dataset.

Summary

We started our first chapter with the data science process that defined the various stages of machine learning. The subsequent chapters covered those stages in-depth. In this final chapter, I have merged all those steps and provide you with definite guidelines and paths to follow to develop the efficient machine learning models. Hope this entire exercise makes you a successful data scientist. Good luck!

Printed in the United States
by Baker & Taylor Publisher Services